D1386518

Hidden Histories
of Science

Edited by
Robert B. Silvers

Granta Books
London

Granta Publications, 2/3 Hanover Yard, London N1 8BE

First published in the USA by the *New York Review of Books*, 1995
First published in Great Britain by Granta Books 1997
This edition published by Granta Books 1998

A CIP catalogue record for this book is available
from the British Library.

1 3 5 7 9 10 8 6 4 2

Printed and bound in Great Britain
by Mackays of Chatham PLC

Contents

Introduction

Late one night after dinner in a New York restaurant Oliver Sacks told me that he had been reading with much excitement a newly published life of Humphry Davy, the great English chemist who in the early nineteenth century had first isolated elements including potassium, sodium, calcium, and magnesium, and had invented, among other things, the safety lamp for coal miners. Davy, he told me, had been his boyhood hero and now, reading the new biography by David Knight, he recalled what he had partly forgotten: that Davy had been a friend of Wordsworth and Southey and particularly of Coleridge, who came to his lectures. Davy wrote poems himself and lived at a time when science and poetry were both seen as imaginative enterprises, complementary ways of exploring the natural world.

As any other editor would have done, I asked Oliver to write on Humphry Davy and the new book about him and within a few weeks he sent us a powerful essay, not only on Davy but on the history of science and how it could harbor neglected and deeply suggestive insights into the workings of the mind and nature.

> Science sometimes sees itself as impersonal, as "pure thought," independent of its historical and human origins. It is often taught as if this were the case. But science is a human enterprise through and through, an organic, evolving, human growth, with sudden spurts and arrests, and strange deviations, too. It grows out of its past, but never outgrows it, any more than we outgrow our own childhood.

Just as we were about to publish this essay the New York Public Library asked *The New York Review* to organize a series of lectures on a single topic. Why not, we thought, take up Oliver's theme of forgotten and neglected moments in the history of scientific discovery? We turned to some of our longstanding contributors on scientific subjects and asked them to consider writing on episodes, or themes, in the history of science that seemed to them worth recalling, not least because of what they suggested about the uses or implications of scientific history itself.

We put our question to some of the scientists and writers on science whose work we have particularly valued for the depth and lucidity of their writing and for their ability to connect scientific discoveries with their historical settings and social consequences. Without saying exactly what they would choose for a subject, they all accepted our invitation. As the months passed, however, we found, without any urging on our part, that some common themes were emerging in the lectures and the essays based on them. Several of our contributors showed how discoveries and insights could emerge with what seemed great promise, and yet be pushed aside, discarded, and forgotten—only to re-emerge once again, sometimes many years later, and become, in their new formulation, accepted as important.

Just how and why this happened is the subject of three of the essays published here—by Jonathan Miller, Oliver Sacks, and Daniel Kevles. The essays by Richard Lewontin and Stephen Jay Gould are not unrelated, suggesting deep and largely unacknowledged distortions in the ways scientists, and popularizers of science, conceive the structure of the world and its natural history.

As Barbara Epstein and I have often found in editing *The New York Review*, the pleasure of editors is in learning, from brilliant thinkers and writers, things we could never have otherwise known or even suspected. So it has been with the essays in this book. It would not have come into being without the help of Barbara Epstein, Rea Hederman, and the staff of *The New York Review*, particularly its Associate Publisher, Catherine Tice, and its Promotion Director, Andrea Barash, and if we had not been asked to present a series of lectures at the New York Public Library. To all of them, and above all to our contributors, my somewhat amazed thanks.

—Robert Silvers

Jonathan Miller

Going Unconscious

I

It was in a mood of irritable skepticism that the Scottish surgeon James Braid attended a public demonstration of Animal Magnetism on the night of November 13, 1841. From everything he had read and heard about the trances that occurred during such demonstrations at the bidding of the operator—the person who induced the trances—he reports that he was "fully inclined to join with those who considered the whole thing to be a system of collusion and delusion, or an excited imagination, sympathy, or imitation." After observing the demonstration, he considered that the trances were quite genuine, but at the same time he felt satisfied "that they were not dependent on any special agency or emanation passing from the body of the operator to that of the patient as animal magnetizers allege." He returned to the demonstration when it was repeated by popular demand a week later and on this occasion he felt that he had identified the cause of these mysteriously punctual onsets of "nervous sleep." He was to devote the last eighteen years of his life to the topic, and under the proprietary title of Hypnotism he explained and redescribed the process in terms which would have been unrecognizable to its eighteenth-century discoverer, Franz Anton Mesmer.

2   Mesmer was born in the lakeside town of Constanz in 1734, and after receiving a philosophical education from the local Jesuits he studied medicine at the University of Vienna and qualified as a physician in 1767 with the publication of an MD thesis on the influence of celestial gravity on human physiology. He argued that the rotation of the heavenly bodies exerted gravitational influence on human physiology analogous to the tidal effect of the moon upon the ocean and that this accounted for the periodic incidence of various diseases. To explain the transmission of this influence he invoked the existence of an immaterial substance, a weightless or so-called imponderable fluid, whose universal distribution guaranteed action at a distance. The notion of such an ethereal medium can be traced back to Greek antiquity and under various titles—ether, spiritus, pneuma, etc.—it figures as a recurrent theme in European scientific thought. It played a prominent part in the long tradition of Neoplatonism and in all probability it was under the influence of this somewhat occult tradition that Isaac Newton cautiously invoked the existence of "a subtle spirit which pervades and lies hid in all gross bodies; by the force and action of which spirit, particles of bodies attract one another." The medium is frequently mentioned in Newtonian unpublished correspondence, and there are several references to it both in the *Principia* and in the more widely read *Optics*.

Although Newton stressed it was no more than a hypothesis, some of the commentators who sought to explain and publicize his work appear to have taken the existence of the Newtonian ether quite literally, insisting that it was the only intelligible explanation for the distant transmission of gravity, light, heat, and magnetism. It was evidently from one of these secondary sources that Mesmer derived the concept of an imponderable fluid and then applied it uncritically to the questionable effect of gravity upon human physiology.

For obvious reasons Mesmer did not foresee any remedial implications of his theory, but on learning that a Jesuit astronomer had conducted successful clinical experiments with magnets, he decided to follow suit, recognizing an affinity between gravity and magnetism.

By drawing powerful magnets over the limbs of patients afflicted with a variety of disabling conditions, Mesmer believed that he was wielding a local influence comparable to that of celestial gravity and that the effect was mediated by the ethereal substance which he had mentioned in his MD thesis.

In the ensuing squabble over priority, Mesmer abandoned the magnetic technique for which he would have had to share the credit, and since he was eager to establish an exclusive claim, he announced that he had identified a previously unrecognized form of magnetism whose application for medical purposes did not require the disputed use of ferrous metals. According to him the human nervous system was charged with "animal" magnetism— so-called because it was associated with the soul or *anima*—

'*Une Séance de Magnétisme de Mesmer, dans son hôtel de la Place Vendôme,*' after a woodcut by Figuier

and although it was not physically detectable he insisted that it could be mobilized by the will of an initiated practitioner and broadcast to living bodies in the vicinity.

He administered the treatment by passing his hands up and down the length of the patient's body without actually touching the surface. In a series of conspicuously uncontrolled trials he reported dramatic improvements, noting that the most favorable outcome was obtained when the patient reacted to the "magnetic" passes by falling into a convulsive trance. In fact, the fits and trances began to assume an emblematic importance and among a significant proportion of his clientele they were valued just as much as the medical relief.

*   *   *

The combination of unorthodox procedures on the part of the practitioner and the frequently unseemly behavior on the part of the patient predictably aroused the suspicion of Mesmer's professional colleagues in Vienna, and in 1778 he felt obliged to emigrate to Paris. In a metropolitan city in which the educated public was almost frivolously susceptible to scientific novelty, a medical treatment based on a previously unacknowledged force of nature proved irresistibly attractive; and since the orthodox profession tormented its clients with ineffectual purges and emetics, the arrival of animal magnetism was heralded as a medical millennium. In any case, the trances and "fits" alone were worth the price of admission, and for many of the clients who attended as spectators the magnetic "crises" were the main attraction. Once again, the incontinent festivities became a subject of scandalous controversy, and in 1785 two official commissions were appointed to investigate the matter. Both of them reported unfavorably, but from the scientific point of view the most damaging conclusions were the ones published under the auspices of a commission appointed by the Academy of Sciences. The commission, which included Benjamin Franklin and Lavoisier, insisted that neither the trances nor the cures had anything to do with magnetism but depended on what would now be called the placebo effect, that is to say, upon the patient's powerful expectation of the forecast result. If the "magnetic" procedure was inaudibly performed behind a magnetically transparent screen, so that the patient was unaware of it, nothing happened, proving that the mesmeric enterprise depended upon the patient's susceptible imagination and not, as Mesmer argued, upon a peculiar form of magnetism emanating from the operator.

It was not the first time that "imagination" had been invoked to explain anomalous medical phenomena. By the end of the eighteenth century there was a large body of specialist literature devoted to the physiological influence of the mind upon the body and although the term "imagination" suggests to the modern reader that the effects were visualized as being unreal or imaginary, for the eighteenth-century theorist it implied that the effects, real enough in themselves, were *caused* by the imagination. *How* they were caused remained a mystery and to that extent the theory of the imagination is scarcely a theory at all. Nevertheless, it had the advantage of being considerably less implausible than the theory of animal magnetism and for that reason "imagination" was repeatedly cited by scientists and physicians who found the rigmarole of imponderable fluids and magnetic emanations philosophically unacceptable.*

Another reason for favoring the otherwise inconclusive theory of the imagination is that by putting the emphasis on the patient's susceptibility it diminished the vainglorious pretensions of the mesmerists and of Mesmer in particular. Although he advertised animal magnetism as a natural function and implied that its remedial potential could be realized by anyone who took the trouble to master it, as time went on, his behavior gave a different impression altogether. He jealously guarded the details of his technique, and from his dress and his demeanor in the magnetic salon, it seems clear that he was happy to project the image of a magus. Clothed in a robe embroidered with Rosicrucian alchemical symbols, he stalked the darkened rooms to the accompaniment of a glass harmonica and actively encouraged his clients to luxuriate in their convulsive crises. By 1785 the whole enterprise had become melodramatically inconsistent with scientific professionalism and the scientific establishment did what they could to close him down.

Although Mesmer's personal reputation suffered irreversibly as a result of these criticisms, the cult of animal magnetism was, if anything, encouraged by the controversy and throughout Europe it continued to attract clients precisely because it was the subject of official criticism. In fact, it was visualized in its own time in much the same way as unorthodox or fringe remedies are today, i.e., as part of a subversive counterculture.

8

*   *   *

A few months after the publication of the French reports, paragraphs appeared in the London press advertising introductory lectures and demonstrations of the "new method for healing all known diseases," and by 1786 mesmerism was flourishing in the hands of local practitioners, at least two of whom claimed to have learned the technique from Mesmer himself.

Although it was advertised as a remedial treatment, the clientele was by no means confined to the sick and disabled. On the contrary, many of those who attended the séances were attracted by the metaphysical implications of the doctrine, by what M. L. Abrams called its "natural supernaturalism." With its intriguing combination of occult powers, clairvoyant trances, and invisible weightless fluids, animal magnetism seemed to guarantee the existence of a reality beyond the world of the senses and many people saw it as an irresistible alternative to an increasingly mechanized picture of the universe.

Political events on the other side of the English Channel soon cast a shadow of official suspicion over anything associated with France, and since animal magnetism had an additional ingredient of subversive radicalism it rapidly lost its following and by 1794 it more or less vanished from the English scene.

The interest was dormant rather than dead, however, because the arrival of a visiting French magnetist—Baron Dupotet—in 1836 set off one of the most controversial episodes in the history of the subject, incidentally wrecking the academic career of one of London's most distinguished physicians.

Shortly before the arrival of the mesmeric Baron Dupotet, John Elliotson, a close friend of Dickens, Thackeray, and Wilkie Collins, had been appointed to the Chair of Medicine at University College, London. Apart from his distinguished contributions to clinical medicine, he was widely regarded as one of the more "philosophical" members of the profession, and in his preoccupation with some of the larger issues of Life and Mind, he often dismayed his colleagues by sponsoring unorthodox causes. For example, he was president and founder of the London Phrenological Society, and since there seemed to be an elective affinity between phrenology and animal magnetism, Elliotson greeted the visiting mesmerist as a kindred spirit. After witnessing his first public lecture he invited him to conduct further demonstrations in his own wards at University College.

It is difficult to imagine a more inappropriate setting for such an experiment. The public wards of a nineteenth-century charity hospital were filled with impressionably dependent patients, and since Elliotson failed to observe any of the precautions taken by the French commissioners forty years earlier, the misleadingly favorable outcome was a foregone conclusion. Confronted by two impressive practitioners, neither of whom could presumably disguise their coercive optimism, the patients dutifully, albeit unconsciously, enacted what was expected of them and the familiar pantomime of trances, fits, and "cures" duly ensued. And since the behavior of any one subject was unavoidably witnessed by all the others in the ward, the remedial influence of the "imagination" grew by geometric progression, thereby guaranteeing unanimous therapeutic success. By the time the itinerant visitor had taken his departure Elliotson had become an evangelical convert to animal magnetism and in the hospital wards under his jurisdiction an unruly atmosphere of mesmeric revivalism prevailed. By now he had identified a group of patients whose susceptibility was such that they could be relied on to produce dramatic evidence in favor of the new treatment, and before long Elliotson was advertising public demonstrations in the lecture theater at the hospital.

Two women—the Okey sisters—emerged as the stars of this mesmeric cabaret and in the course of the next few years they acquired a legendary reputation as spectacular trance subjects. By all accounts they seem to have been convulsive hysterics, and although they were admitted to the hospital under that label they had unconsciously exploited the same symptoms in a different institution altogether. In his medical memoirs the then assistant editor of *The Lancet* reports that the sisters had already achieved considerable notoriety in a Pentecostal congregation in a nearby church, where their glossolalic interventions had attracted admiring attention. The career of these two young women neatly illustrates the way in which the symptoms of serious personality disorders can be shaped and then reshaped, depending on the social institution in which they manifest themselves. In a congregation which recognized and valued the notion of "speaking in tongues" the sisters modulated their conduct until they were recognizable as Pentecostal prophets, whereas in the wards of the newly converted professor of medicine their repertoire changed under the influence of Elliotson's positive conditioning and they re-emerged as mesmeric shamans.

The scandalous high jinks had by now reached the point where the College Council could no longer ignore them and Elliotson was firmly requested to cease these unseemly demonstrations. He indignantly refused, claiming that he and Mesmer had identified a new force in nature and that by demonstrating its influence to the educated public he was offering an opportunity to exploit a mighty engine for the regeneration of humanity "comparable in importance and power to that of the steam engine." The authorities, however, were adamant and Elliotson resigned in a high dudgeon. He continued to have a successful private practice, however, and since he was no longer tied down by the bureaucratic constraints of the university, he maintained a magnetic practice in parallel to one of conventional medicine. A few years after his resignation he inaugurated, edited, and was one of the leading contributors to a journal entitled *The Zoist*, devoted to the twin topics of phrenology and mesmerism. Naturally, much of the editorial matter was taken up with reports of successful magnetic treatments together with phrenological accounts of notable crania. The journal, however, was characterized by a conspicuous commitment to liberal causes—penal reform, the abolition of capital punishment, the education of workingmen, etc.

Although he was robbed of his prestigious chair, Elliotson continued to enjoy a prosperous professional life with a wide circle of admiring and affectionate friends. Until the early 1850s his name was almost synonymous with that of British mesmerism, and throughout his active life he continued to sponsor the original "magnetic" interpretation of the phenomena.

II

By the time James Braid began his inquiries in Manchester in the early 1840s there was an effective stalemate between those who sponsored the magnetic fluid and skeptics who preferred to explain the phenomena in terms of the patient's susceptible imagination.

As a result of the observations which he made on the occasion of his second visit to the demonstration of animal magnetism, in 1841, Braid effectively broke the deadlock, since he was able for the first time to interpret the subject's contribution in a somewhat more intelligible way.

Braid noticed that the entranced subject was invariably unable to open his eyes, which led him to the belief that the trance was induced by something which he described as the neuromuscular exhaustion induced by the protracted stare encouraged by the operator. To confirm this hunch he invited a friend at home later to gaze unblinkingly at the top of a wine bottle, as he said, "so much above him as to produce a considerable strain on the oculomotor muscles." He described the results in the following words: "In three minutes his eyelids closed, his head drooped, his face was slightly convulsed. He gave a groan and instantly fell into a profound sleep, the respiration becoming slow, deep and stertorous. At the same time his right hand and arm were agitated by slight convulsive movements."

He repeated these experiments on his wife and on his manservant, and having obtained the same results on each occasion, he was convinced that he had efficiently demystified the mesmeric process. He already knew that it had nothing whatever to do with magnetism and the explanation he provided was considerably more specific than the traditional theory of the imagination. All the phenomena, he insisted, were to be explained as the neurological consequences of "a fixed stare, absolute repose of the body, fixed attention, and a suppressed respiration concomitant with that fixity of attention."

By formulating the concept of "nervous sleep," or hypnotism, Braid dealt a death blow to the pseudoscientific theory of animal magnetism and at the same time supplied a more convincing alternative theory. Nevertheless, it is easy to overestimate his intellectual achievement, or to put it more accurately, it is easy to *under*estimate the extent to which Braid was disabled by his own prejudices. The problem is that when someone coins a term by which a concept is currently known, it is tempting to imagine that his sense of the term corresponds to our own and that, like a Yankee at the court of King Arthur,

Braid was a modern psychologist living unrecognized before his time.

The truth of the matter is that although he rendered an important service by stressing the condition of the patient's nervous system as opposed to supernatural power on the part of the practitioner, in almost every respect he turns out to have been just as credulous as the mesmerists that he criticized.

For example, as soon as he discovered how easy it was to induce the hypnotic trance by fixed attention, looking at bottle tops and so forth, he visualized almost unlimited prospects of therapeutic effectiveness. Nearly two thirds of the first book which earned him his fame is taken up with implausible accounts of dramatic cures achieved under the influence of "nervous sleep."

The skepticism with which he originally approached the subject seems to have gone out of the window. Here, he records his experience with a congenitally deaf patient.

Hitherto, these patients have been considered beyond the pale of human aid. The morbid condition of the organs as ascertained by dissection was sufficient to warrant the inference that it was improbable that any remedy could be discovered for such cases.

Fully aware of this pathological difficulty I was nevertheless inclined to try the effect of neurohypnotism with congenital deaf mutes, knowing that it could be done with perfect safety, without pain or inconvenience to the patients.

From having witnessed its extraordinary power of arousing the excitability of the auditory nerve, I entertained the hope that it might be capable of exciting some degree of hearing from the increased sensibility of the nerves compensating for the physical

imperfection of the organ. The result of my first trial
was beyond my most sanguine expectations which induced me to persevere and the result has been that I
· have scarcely met with a single case of the congenitally
deaf mute where I have not succeeded in making the
patient hear to some degree.

With undiscriminating enthusiasm, Braid records
more than fifty cases, ranging from hemiplegias, spinal
curvatures, to ankylosing spondylitis and epilepsy, and in
almost every instance, he reports conspicuous improvement. Even the most ardent supporter of hypnotism
would find it hard to identify with Braid's claims.

In fact, as one reads on, it begins to look as if the
imagination of the physician is just as relevant to the outcome. Certainly this seems the case when it comes to
some of the other phenomena which hypnotism supposedly brought to light. Braid, it seems, like his colleague
Elliotson in London, was an ardent phrenologist. And
as soon as he learned to control the hypnotic condition,
he was exploiting it to prove the existence of the cerebral
organs which had previously been identified by Gall and
Spurzheim.

By touching the hypnotized patient at various points
on her skull, he was gratified to witness behavior that corresponded to the function of the organ which supposedly
lay beneath the magnetizer's finger.

Miss S, who knew nothing of either hypnotism or
phrenology, sat down and in a few minutes she was
not only decidedly hypnotized, but was also one of the
most beautiful examples of the phrenological sway
during hypnotism.

> The moment the organ of veneration was touched, her features assumed the peculiar expression of that feeling. Her hands were clasped. She sank on her knees in the attitude of pious adoration. When the organ of hope was touched in addition, the features were illuminated and she beamed with a feeling of ecstasy. On changing the points of contact to the organ of Firmness, she instantly arose and stood in an attitude of defiance.
>
> A touch on the organ of self esteem and she flounced about the room with the utmost self importance. Under Acquisitiveness, she stole unashamedly, and yet when Conscientiousness was stimulated, she instantly restored the stolen property.
>
> Her philoprogenitiveness was admirable.

And so on, page after page. The results are so bizarre and so self-evidently questionable that one is struck by the disproportionate pedantry with which Braid sought neurological explanations for what had occurred.

He recognized, for example, that the scalp was innervated—i.e., anatomically supplied—by the fifth cranial nerve, and conceded that none of its fibers passed directly through the skull to the brain beneath. Nevertheless, according to him, there was every reason to believe

> that the distribution of the nerves of the scalp will be ultimately found far more intricate and beautifully arranged than at present we have conception of. And that the cerebral extremity of each fiber will be found to be connected with the peripheral extremities of a single fiber only. And that this peripheral extremity is in relation with only one point in the brain or spinal cord.

Ingenious though this baroque scheme is—and one might want to say it foreshadows modern concepts of mapping the body surface onto the cerebral convolutions of the brain—it is absolutely absurd as an explanation of the improbable behavior cited by Braid.

And yet, as it happens, at least two of Braid's contemporaries spotted something indispensably valuable in all this phrenohypnotic tomfoolery. But what they saw had nothing to do with either phrenology or remedy.

Braid's "hypnosis," now stripped of its magnetic pretensions, seems to have revealed to these observers something interesting about the functions of the nervous system. Something for which there was plenty of anecdotal evidence but, as yet, no repeatably experimental facts.

Benjamin Carpenter, one of the most distinguished physiologists of the nineteenth century—a professor at University College in London—insisted that Braid's research had indeed thrown more light on what he, Carpenter, described as the reflex functions of the brain than any other investigations hitherto. So also did Thomas Laycock, professor of medicine at Edinburgh and the teacher of the neurologist John Hughlings Jackson at York—although Laycock was somewhat more preoccupied with proving that it was he and not Carpenter who first recognized the existence of reflex action above the spinal cord.

III

But what did both of these men mean when they referred to the reflex function of the brain, and how had Braid's work helped to elucidate it? What they were referring to is that vaguely defined area of both action and cognition, which lies between the unarguably automatic and the self-evidently voluntary. Or to put it another way, between actions of which the individual has no conscious control, and actions and cognitive processes for which the existence of consciousness is absolutely essential.

At the same time it was generally recognized, and not only by physiologists, that between these two apparently distinct provinces of human behavior, there was a broad strip of territory in which it was not all that easy to determine the exact identity of the behavior or the cognition in question.

For example we all know that when the ground underfoot is treacherous and uncertain, we have to devote all our attention to the act of walking. And yet when the going is good and our attention is diverted by an interesting conversation, we can stroll along, confident that the strides will take care of themselves. Learning to play a piece of music will monopolize all our attention while we're learning, but once we've mastered the fingering, we are capable of playing the melody and devoting our thoughts elsewhere — perhaps to imagining how large a fee we can demand for the performance.

The same principle applies to perception. We all know that heedlessness is more apparent than real and that as long as we are prompted or "cued" in the right way, we're often surprised to recall seeing something while our attention was apparently elsewhere. Everyone knows the experience, the alarming experience, of arriving home with a problem on his mind, not having remembered a single episode on the journey, until reminded of it. Laycock and Carpenter gave special emphasis to involuntary memory, the well-known experience of effortlessly recalling a forgotten name after hours spent strenuously trying to remember it. And Carpenter records in his books many examples of solving intractable problems just by sleeping on them. Laycock and Carpenter insisted that these facts revealed something fundamental about the way in which mental activity was organized. Both men quote the same passage from the philosopher Sir William Hamilton in support of this claim. This is what Hamilton wrote in 1842:

> What we are conscious of is constructed out of what we are not conscious of. Our whole knowledge in fact is made up of the unknown and of the incognizable. There are many things which we neither know nor can know in themselves, but which manifest themselves indirectly through the medium of their cognitive effects. We are thus constrained to admit, as modifications of mind, what are not phenomena of consciousness.

Far from claiming any originality for this insight, Sir William Hamilton argued that the credit was largely due to Leibniz, while indicating at the same time that Leibniz had been unfortunate in the terms that he employed to propound his doctrine. By using such terms as "obscure representations," "insensible ideas," or "petites perceptions," Leibniz was, according to Hamilton, violating the universal characteristics of language. "For perception, idea and representation," he said, "all properly involve the notion of consciousness." Not necessarily, argued John Stuart Mill. In his *Examination of Sir William Hamilton's Philosophy*, he pointed out there was another way of resolving this apparent contradiction. Mill writes:

> If we admit what physiology is rendering more and more probable, that our mental feelings as well as our sensations, have for their physical antecedents, particular states of our nerves, it may well be believed that the apparently suppressed links in a chain of associations, really are so. So that they are not even momentarily felt. The chain of causation being continued only physically by one organic state of the nerves succeeding so rapidly, that the state of consciousness appropriate to each is not produced.

I'm not suggesting that either Laycock or Carpenter was directly inspired by either Hamilton or Mill, though each quotes both philosophers. But I think that Hamilton and Mill encouraged Laycock and Carpenter to press for the recognition of a cerebral process analogous to the automatic reflexes of the spinal cord. Carpenter called it *Unconscious Cerebration*, whereas Laycock stuck to the phrase with which he claimed to have originated the idea, i.e., "reflex function of the brain."

The reason both Carpenter and Laycock stress the importance of Braid's work, notwithstanding its phrenological excesses, is that the hypnotic trance, as they saw it, exposed the action of these unconscious processes. By artificially paralyzing the will, a broad layer of automatic action was now made conveniently visible, or so they claimed.

The truth is that it's hard to tell just how much Braid's contribution weighed in the final outcome. Laycock had published his first intimations on the subject of the reflex functions of the brain a year before Braid witnessed the magnetic exhibition in Manchester. And although Carpenter refers with increasing frequency to the significance of hypnosis in the subsequent editions of his *General Physiology*, there were preexisting reasons for believing that the neurological function of the brain was not confined to either consciousness or voluntary action and that in spite of all superficial appearances to the contrary, it retained a large repertoire of automatic actions.

The most persuasive argument in favor of this idea was one to which Thomas Laycock had been exposed during his student years in Germany. In subsequent publications he acknowledged the influence of a group of theoretical biologists known as *Naturphilosophes* who insisted that the design of the vertebrate nervous system was inherited from an ancestral prototype in which a set of homologous modules was arranged in a linear series from one end of the body to the other. According to the theory, the central nervous system of this admittedly hypothetical ancestor consisted of a sequence of identical segments and there was, as yet, no visible distinction between the segments at the front end of the body and those at the rear. But with the development of senses specialized to detect distant events—eyes, ears, and olfactory organs—a process called cephalization began to set in and a recognizable head started to make its appearance. The first ten segments of the central nervous system then enlarged and coalesced to form a brain, distinguishable from the spinal cord which retained the ancestral pattern of serial segmentation. According to the theory, neurological function was correspondingly differentiated. The spinal cord continued to execute automatic reflexes in response to local stimuli, while the brain, now in receipt of information about more remote events, assumed overall control and acted as a command module.

But although the brain increasingly undertook the responsibility for more thoughtful behavior, the fact that it was derived from neural segments or ganglia which were once indistinguishable from the ones that composed the spinal cord meant that it preserved a discoverable repertoire of automatic actions. It was under the influence of this theory that Laycock addressed the York meeting of the British Association for the Advancement of Science in 1844.

> Four years have lapsed since I published the opinion that the brain, although the organ of consciousness, is subject to the laws of reflex action. I was led to this opinion by the general principle that the ganglia within the cranium, being a continuation of the spinal cord, must necessarily be regulated as to their reaction to extended agencies by laws identical to those which govern spinal ganglia and their analogues in lower animals.

A careful reading of Laycock's *Mind and Brain*, written in 1860, shortly before he died, shows how much he owed to the German *Naturphilosophes*. But he was also influenced by the mesmeric evidence. In an unpublished manuscript now in the Edinburgh University Library, Laycock narrates his earlier experiences in London when he encountered the bizarre automatisms of Elliotson's notorious Okey sisters. Primed by his earlier exposure to German philosophical biology, the mesmeric evidence inaugurated what was to be a lifelong interest in the shadowy province between two kinds of behavior: the unarguably automatic and self-evidently voluntary.

However, as far as orthodox neurological opinion was concerned there was no such province. On the contrary, the functions of the brain were not to be confused with those of the spinal cord. For example, according to Marshall Hall, one of the great pioneers of English neurophysiology,

> The functions of the *cerebral* system [i.e., the brain] are sensation, perception, judgement, volition and voluntary motion. The cerebrum itself may be viewed as the organ of mind. It is the organ on which the psyche sits, as it were, enthroned. All its functions are strictly psychical. They imply consciousness. Sensation without consciousness appears to one to be a contradiction in terms. How different from those which I have just enumerated are the functions which belong to the true spinal nervous system. In these there is no sensation, no volition, no consciousness, nothing psychical at all.

Hall's conclusions were based on animal experiments in which he destroyed the brain, leaving the spinal cord isolated but intact. In the animals which survived this maneuver a large repertoire of muscular reactions were preserved. Whereas more complex, volitional behavior disappeared altogether.

But it was not the experimental evidence alone that prompted Hall to make such a hard and fast distinction between the spinal cord on one hand and the brain on the other. As a pious Christian who carried a Bible wherever he went, he was eager to establish a neural province within which the immortal soul enjoyed unquestioned sovereignty beyond the reach of profane materialism. Like Descartes, almost two hundred years earlier, Hall was prepared to make a large territorial concession to mechanism in exchange for a treaty which recognized the local sovereignty of the soul and the brain. The only difference was that whereas Descartes' soul was confined to the somewhat cramped premises of the pineal gland, Hall furnished the spiritual monarch with the large upholstered apartments of the brain as a whole.

A hundred years earlier, La Mettrie had violated the Cartesian treaty by declaring that Man was a machine through and through, indistinguishable from the automata created by the Swiss toymaker Vaucanson. And that although this imprudent Anschluss had been resisted, it was important to police the cerebrospinal frontier. So that when Laycock and Carpenter surrendered a large part of intracranial territory to reflex activity, there were those who understandably felt the free will and sovereignty and, indeed, the divinity of man were in jeopardy.

Carpenter, however, who was just as pious as Marshall Hall, had no misgivings at all when it came to conceding the existence of something he called Unconscious Cerebration, which allowed mechanism to creep upward into the cranium itself. For one thing, the evidence in favor of it was by now incontestable, especially since Braid's work had thrown such a strong light on it. And anyway, for Carpenter, the Will, being an unextended agency, could afford to lose a few ganglia to the encroachments of mechanism. Indeed in a concession which was the neurological equivalent of Henri IV's conceding that Paris was worth a Mass, Carpenter enlarged the province of the automatic to secure the sovereignty of the will.

Paradoxically, it was in the face of yet another mechanistic onslaught that Carpenter reaffirmed his argument in favor of the reflex function of the brain. In the preface to the fourth edition of his *Human Physiology*, he gave a summary of T. H. Huxley's recent address to the British Association in Dublin.

Huxley had argued, in an article now entitled "Animals Are Automata," that Man himself is only a more complicated and variously endowed automaton. Huxley insisted:

> The feeling we call volition is not the *cause* of the voluntary act, but simply the symbol in consciousness of that stage of the brain which is the immediate cause of the act. Like the steam whistle which signals but doesn't cause the starting of the locomotive.

To support this claim, Huxley referred to many human automatisms, including the ones obtained under hypnotism. But *that*, according to Carpenter, was the point. The fact that hypnosis reveals so much in the way of automatism proved to him just how important the will was.

As Carpenter understood it, hypnotism induced a temporary suspension of the will which then became all the more conspicuous by its absence:

> The actions of our minds insofar as they are carried without any interference from our will may indeed be

considered as functions of the brain. On the other hand, in the control which the will can exert over the direction of our thoughts and over the motive force of the feelings, we have evidence of a new independent power which may either oppose or concur with the automatic tendencies and which accordingly as it is habitually exerted, tends to render the ego a free agent.

26

In contrast to Marshall Hall, who felt that he had to limit automatism to the, as it were, below-stairs part of the nervous system—the spinal cord—to secure recognition of the spiritual independence of the will, Carpenter was convinced that he could achieve the same theological result at the comparatively modest expense of conceding the existence of automatic functions in the brain and not just in the spinal cord. It would be a mistake, though, to suggest that Carpenter merely conceded the existence of unconscious cerebration. For him, this process represented a function of considerable interest in its own right. So far as voluntary movement was concerned, it showed him how much automatic coordination was involved in executing a conscious act. Here is how he put it in 1876:

> Even in the most purely volitional movements, the will does not directly produce the result. The will plays, as it were, upon the automatic apparatus by which the requisite neuromuscular combination is "brought into action."
> In each of these acts the coordination of a large number of muscular movements is required. So complex are their combinations that the professed anatomist would be unable to say exactly what is the precise state of each of the muscles concerned in the production of a given musical note or in the enunciation of a particular syllable. We simply conceive the tone or syllable we

wish to utter and say to our automatic self, do this and the well trained automaton executes it. What we will is not to throw this or that muscle into action, but to produce a certain preconceived result.

This passage is virtually a paraphrase of a doctrine made famous by the great English neurologist John Hughlings Jackson, who argued that the cerebral cortex represents *movements* and not *muscles*, i.e., that we conceive an *idea*, and throw the automatic self into action.

Now, under normal conditions, this automatic self, unconscious and reflex in character, is at the exclusive disposal of the will. But as Carpenter argues in the 1852 edition of his *Human Physiology*, "In those states in which the directing power of the will is suspended, hypnosis being one of them, the course of action is determined by some dominant idea which, for the moment, has full possession of the mind."

In other words, as the subject's will weakened under the influence of hypnosis, the automatic apparatus of the brain, untouched by the hypnotic effect, is at the disposal of the operator's will, and he is now in a position to dictate both the actions and the perceptions of the enhanced subject.

Neither Carpenter nor Laycock expressed any interest in the neurological mechanism whereby hypnosis brought about this submissive condition. As far as they were concerned, hypnosis, however it worked, was a conveniently reversible technique for diminishing the sovereignty of consciousness and exposing what Hamilton and Mill had previously suspected, i.e., that human beings owe a surprisingly large proportion of their cognitive and behavioral capacities to the existence of an "automatic self" of which they have no conscious knowledge and over which they have little voluntary control.

The role of hypnosis in developing this distinctively *enabling* view of the Unconscious has been regrettably overshadowed by its contribution to the more widely recognized Freudian Unconscious. In fact the modern notion of the Unconscious is so closely identified with the one that figures in psychoanalytic theory that whatever celebrity Mesmer and his successors now enjoy is almost entirely the result of their being seen as antecedents of Freud.

Although there are self-evident points of resemblance, the notion of the Unconscious developed by British psychologists in the middle of the nineteenth century differs significantly from the one advanced by Freud at the end.

In psychoanalytic theory, the Unconscious exercises an almost exclusively withholding function, actively denying its mental contents their access to awareness. Through the agency of repression, which Freud identifies as society's censorious representative in the psyche, the individual is relieved of thoughts which might, if consciously experienced, compromise wholehearted co-operation in social life.

In contrast to this distinctively custodial interpretation, the Unconscious postulated by Hamilton, Laycock, and Carpenter figures as an altogether productive institution, actively generating the processes which are integral to memory, perception, and behavior. Its contents are inaccessible not, as in psychoanalytic theory, because they are held as in strenuously preventive detention but, more interestingly, because the effective implementation of cognition and conduct does not actually *require* comprehensive awareness. On the contrary, if consciousness is to implement the psychological tasks for which it is best fitted, it is expedient to assign a large proportion of psychic activity to automatic control: if the situation calls for a high-level managerial decision, the Unconscious will freely deliver the necessary information to awareness.

When represented in these terms, the Unconscious visualized by Carpenter and Laycock anticipates its role in late-twentieth-century psychology and I suspect that if the long drought of Behaviorism had not taken place when it did, it would not have required an intellectual revolution to inaugurate the age of cognitive psychology.

As it is, though, by the end of the nineteenth century, the scientific community had become increasingly unfriendly toward explanations which appealed to mental processes that were publicly unobservable. Consciousness itself was bad enough, since it was, by definition, incorrigibly private, so the notion of a mental process, which was subjectively inaccessible into the bargain, struck many psychologists as perverse and unhelpful. The result was that the concept of Mind itself underwent an eclipse and the experimental study of "behavior" replaced it. Psychologists devoted themselves to the study of the relationship between measurable stimuli and quantifiable responses, and preserved their scientific chastity by abstaining from speculations about the mental processes which intervened.

The new regime was accordingly christened Behaviorism and for the next thirty years explanatory references to the "Mind" were regarded as academically suspect. Such discourse was confined to the scientifically segregated community of psychotherapy with the result that the notion of the Unconscious acquired almost exclusively Freudian connotations and the role of hypnotism in promoting the discovery of the alternative version was forgotten.

The revival of the alternative, non-Freudian Unconscious began in the 1950s when its reconstruction under the auspices of artificial intelligence was one of the factors responsible for the timely decline of Behaviorism. As George Miller put it, the Mind returned to scientific psychology, legitimized by its metaphorical identification with the computer.

Research into automatic aiming devices during the Second World War had shown that the control mechanism needed to pursue and destroy an unpredictably moving target required an internal representation in which to register, update, and compute the relevant information. Although no one claimed that such an internal representation was conscious of its own deliberations, the necessity of such a mechanism gave credence to the notion of unconscious information processing and the so-called cognitive revolution ensued as a result.

Reconstructed in computational terms, the enabling unconscious, whose existence had been anticipated by

Mill, Laycock, and Carpenter, began to figure mostly in psychological discourse.

*   *   *

One of the earliest expressions of this fruitful revival is to be found in the work of Noam Chomsky, who stipulated the necessity of covert mental activity to account for the distinctive creativity of language. In an epoch-making criticism of Behaviorism, Chomsky argued that it was impossible to explain our ability to utter and understand sentences that had never previously been spoken or heard without invoking the existence of a linguistic Unconscious capable of generating such versatile competence. He also drew attention to the deep syntactic resemblance between widely different natural languages and insisted that structural similarities required an abstract representation common to all of them. Although the details of this universal grammar were subjectively inaccessible, he argued, it consistently supplied the wherewithal for the rich diversity of conscious communication.

The revolution in linguistics has had its counterpart in almost every other department of psychology as Behaviorism has retreated in the face of evidence favoring the existence of what Carpenter had quaintly described as *Unconscious Cerebration*.

The remarkable phenomenon of "blindsight," for example, bears witness to mental activity of which the individual has no explicit awareness. In patients blinded by injury to the visual cortex, Lawrence Weiskrantz and others were able to demonstrate behavior indicating that the subject had registered the occurrence of a visual stimulus without being subjectively aware of it. The patient was instructed to point toward small points of light unpredictably displayed within the "blind" sector of his visual field. Although he was reluctant to react to an "invisible" stimulus, when he was forced to *guess* where it might be, the accuracy of his pointing was significantly greater than chance. The fear that such a discriminating performance can be achieved in the apparent absence of visual experience indicates a perceptual competence operating well below the level of overt consciousness.

Comparable results have been reported in patients whose brain damage has robbed them of the ability to put names to familiar faces. Such patients have intact language and normal eyesight so that they can correctly identify and name anything else they are shown. Their relatively discrete failure proves that a visual module specifically dedicated to facial discrimination has been destroyed. And yet further experiments indicate that this is not the whole story. When such patients are subjected to a variant of the forensic lie detector test, i.e., when their sweating is monitored by changes in skin conductivity, their reactions surpass chance whenever the photo of a well-loved face is thrown onto the screen. Although they are unable to consciously identify the familiar features, their emotional response proves that they have registered its familiarity, albeit unconsciously.

Experimental work with severe amnesia has yielded analogous results. Patients who consistently failed to

identify test items to which they had been exposed
some time earlier showed that the experience had unconsciously primed their attention to favor items related by
meaning to the ones they seemed otherwise to have forgotten. Let's say, for example, they had been shown a set
of drawings representing musical instruments and that
they failed to recognize any of them as having been seen
before. When shown yet another set of drawings, one of
which represented a musical instrument which they had
not seen previously, they consistently favored that as
opposed to drawings of unrelated objects.

And then there are Edoardo Bisiach's remarkable observations of patients with a unilateral tendency to neglect
or overlook items in their visual field; his findings showed
that it is possible to possess and exploit a spatial representation of the imagined visual world without being consciously aware of it. The patients in question had suffered
local damage to the right parietal lobe of their brain and as
a result they showed a tendency to neglect or overlook
items displayed in the left visual field. Unlike Weiskrantz's
patient they were not actually blind in the affected field,
but simply inattentive, since they would acknowledge
and identify an object in it if their attention was forcibly
drawn to it. Bisiach invited one such patient to imagine
the view of the Piazza del Duomo in Milan, which was
utterly familiar to him by sight before the onset of his ill-
ness. He asked him to visualize and describe everything
that could be seen as viewed from the cathedral steps.
The patient described only one half of what there was to
be seen, insisting, nevertheless, that his description of the

piazza was complete. When invited to list what could be "seen" when imagining the view from the *opposite* side of the piazza, i.e., when facing the cathedral, the patient fluently reported items which he had previously overlooked and claimed no knowledge of whatever.

Experimental results from an ever-widening range of psychological functions tell the same story, that what we are conscious of is a relatively small proportion of what we know and that we are the unwitting beneficiaries of a mind that is, in a sense, only partly our own.

The irony is, it has taken us this long to appreciate what some scientists were telling us more than a hundred years earlier. Yet another example of unilateral neglect!

Notes

* None of the scientists who sat on the Royal Commissions objected to ethereal substances as such. On the contrary, it would have been difficult to find one who did not entertain the notion as **35** part of his world picture. Franklin regarded electricity as a weightless fluid and Lavoisier took it for granted that heat was another one, and until the end of the eighteenth century most scientists accepted the theory that the will transmitted its initiatives to the muscles by setting up vibrations in an ethereal substance closeted in the hollow channels of the nervous system.

Stephen Jay Gould

Ladders and Cones: Constraining Evolution by Canonical Icons

I. Cultures of Presentation and the Role of Iconography

Since the closing years of this millennium will be marked by growing concern and respect for ethnic and cultural pluralism—"the separate and equal station to which the laws of nature and of nature's God entitle them"—"a decent respect to the opinions of mankind" (to continue the quotation from a dead white American male) should also impel us to chronicle the striking differences among our academic communities. The disparities in content between arts and sciences have been much discussed and lamented, most notably by C. P. Snow in his *Two Cultures*, but few have documented, or even mentioned, the striking differences (by no means trivial or superficial) in style and manner of presentation.

Consider, for example, the important academic forum of oral presentations at professional meetings, an essential launching pad for nearly all scholarly careers. The two major differences between scientific and humanistic styles of presentation strike me as wondrously ironic. In stereotypes well known to all, scientific talks may possess empirical content, but usually fail for want of linguistic grace or skill in communication, while humanists, at their best, will at least dazzle with thoughts "ne'er so well expressed," even if the ten thousandth analysis of Shakespeare's one hundredth sonnet fails to present anything truly novel in content. Yet—and hence my judgment of irony—the two major differences between professions show superior intuition among scientists about use of language and style of communication.

First, humanists almost invariably read their papers from a written manuscript (and almost always badly, with head buried in text and bland inflexion quite unsuited for oral presentation). Scientists hardly ever read; we think through the order or logic of the argument, make outlines and notes, and then speak extemporaneously. I would have thought that the superiority of such truly oral presentation would be self-evident. First of all, as a practical matter, the scientists' strategy takes so much less time for the same amount of genuine care (many of the humanists' written documents are not meant for later publication and truly represent wastage).

Second, extemporaneous speech is so much more attractive and compelling of attention than the bland and spiritless style of most readers. I realize, of course, that a good reader can overcome this obstacle with a few simple rules (like memorizing a sentence at a time and looking up at the audience), but, in practice, few people read well—

and the aggregate boredom of bad reading far outweighs the cumulative awkwardness of dubious grammar and parsing among scholars unskilled in extemporaneous speech. I assume, by the way, that many humanists adopt a strategy of reading out of fear, for linguistic style is their *summum bonum*, and they will trade freedom from one spontaneous misconjugation for overall tediousness and even incomprehension—while scientists, who are not much judged by their peers for linguistic style, will opt for better communication with a few potential errors.

But thirdly, and most importantly, written and spoken English are utterly different languages—and humanists, above all, should know this. Documents meant for speaking usually don't work in print (Martin Luther King's "I have a dream" is the greatest speech of the twentieth century, but, as oral poetry based on rhythmic repetition, it reads terribly). The differences are legion. To cite just one: oral speech needs a cyclic structure of studied repetition, for presentation is linear and the listening audience cannot go back; but written documents may be more sequential and non-redundant because a reader can pause to consult an earlier passage. (I have found over the years that the redundancy in a good extemporaneous speech sets the major reason for such discouragement when one reads a transcript. "Did I say such drivel?" one wonders—but the presentation *was* good.)

As a second difference between talks of scientists and humanists, scientists nearly always show slides (or visual material in some other form), while humanists usually rely on text alone (with some striking and obvious exceptions, like art history, where simultaneous use of two slide projectors has become *de rigueur*). Slide projectors are always and automatically provided for any scientific talk. I never think about asking for one; I simply assume that the machine will be there. Therefore, I frequently forget to submit the specific request that must be made when giving a speech to humanists (often in rooms with no screens and no way to darken windows). On three embarrassing occasions, I have shown up with a talk to humanists absolutely dependent upon slides only to find no means for projection. In each case, I was able to put forth an SOS for projector and screen to a colleague in a scientific department.

This striking difference even applies to talks by humanists about explicitly visual subjects. I recently attended a conference in Paris to celebrate the two hundredth anniversary of the Musée d'Histoire Naturelle. Talk after talk commemorated the great scientists (Cuvier and Lamarck, for example) and spoke in detail about the exhibits and the importance of their arrangement and aesthetics, but almost no one showed a picture.

Why do scientists grasp the importance of visual imagery, while most humanists accept the hegemony of the word? Scholarly publication in the humanities generally degrades imagery and in many ways. Many thick tomes have no pictures at all—not even a likeness of a central figure in a narrative. Images, when present, are often only "illustrative" in the slightly pejorative and peripheral sense; they are often collected in separate sections, divorced from textual reference and therefore subsidiary.

But visual imagery is central to our lives. Speaking biologically, primates are the quintessentially visual animals among mammals (a glance at the standard "homunculus" image of the human brain shows how much of the cerebral cortex serves our visual system). Much of our judgment in social matters, particularly our emotional feelings, depends upon images. Where would patriotism be without the Statue of Liberty, the Spirit of '76, and the raising of the flag on Mount Surabachi. And try understanding modern American culture without Ms. Monroe over the subway grating or Mr. DiMaggio at the bat.

From a scholar's point of view, much can be learned from the study of imagery (including its neglect). Since humanists take words as their explicit stock-in-trade, they scrutinize texts with intense care and invest most of their attention in removing biases and clarifying arguments. Since iconography is usually seen as superfluous, motives that attend the choice and form of images are less conscious than those of scientists—and therefore underlying personal and social biases become exposed in the pictures that we use.

I am particularly intrigued by the subject of "canonical icons," i.e., the standard imagery attached to key concepts of our social and intellectual lives. Nothing is more unconscious, and therefore more influential through its subliminal effect, than a standard and widely used picture for a subject that could, in theory, be rendered visually in a hundred different ways, some with strikingly different philosophical implications. The shock of seeing non-standard imagery can be revealing: we instantly realize how constraining the canonical icon had been, though the limitation had never before crossed our mind. For example, as a Jew with no great stake in the subject, I was struck by how unnerving I found the beardless Jesus of Byzantine imagery when I first saw this representation (and realized that we knew absolutely nothing about the appearance, not to mention very little about the existence, of the historical Jesus).

* * *

This essay treats the canonical imagery of my own profession: evolution and the history of life. I know no other subject so distorted by canonical icons: the image we see reflects social preferences and psychological hopes, rather than paleontological data or Darwinian theory. This theme of constraint by standard pictures is particularly important in science, where nearly every major theory has a characteristic icon. Consider the standard rendering of the Copernican solar system (or the Keplerian version with corrected orbits), and then recognize how much the Bohr atom became the microcosm of this macrocosmic icon. The Cartesian geometry of the celestial icon may be empirically adequate, but drawing electrons as planets cycling about the neutrons and protons of a central "sun" does not accurately represent the atomic world.

II. The Ladder or Linear March of Evolution

The most serious and pervasive of all misconceptions about evolution equates the concept with some notion of progress, usually inherent and predictable, and leading to a human pinnacle. Yet neither evolutionary theory nor life's actual fossil record supports such an idea. Darwinian natural selection only produces adaptation to changing local environments, not any global scheme of progress. We can interpret local adaptation as "improvement" in a particular circumstance (the hairier elephant that becomes a woolly mammoth does better in ice age climates), but a historical chain of sequential local adaptations does not accumulate to a story of continuous progress. (The vector of climatic change is effectively random through time, so why should creatures, tracking such vectors by natural selection, become better in any general sense?) Moreover, for each local adaptation achieved through increasing complexity by some definition, another equally successful local solution evolves by "degeneration" of morphology or behavior. (Consider only the numerous parasites that, protected from the rigors of external environments, become little more than bags of feeding and reproductive tissue attached to the bodies of their hosts; yet the parasites have as much prospect of evolutionary success as the hosts.)

As for the fossil record, its pattern of nearly three billion years of exclusively unicellular life, followed by the introduction of nearly all major multicellular groups in a single episode lasting some five million years (the famous "Cambrian explosion" of 535–530 million years ago), grants little credence to any idea of slow and steady advance. At the very most, one might say that a few lineages have expanded into the originally empty sphere of anatomical elaboration (since life had to arise at the lower limit of its conceivable, preservable complexity—that is, as tiny, simple, single cells). But, without question, these earliest and simplest cells, the bacteria and their allies, remain the most abundant, widespread, and successful of all living things. And if one insists on multicellular animal species, some 80 percent of them are insects, and these enormously successful creatures have not shown any pervasive vectors of improvement over the past 300 million years.

This conceptual problem has pervaded evolutionary biology ever since Darwin. The very word "evolution," as a description of biological change through time, entered our lexicon through Herbert Spencer's more general usage (for cosmology, economics, and a host of other historical disciplines) in the service of his firm belief in "universal progress, its law and cause." Darwin himself had consciously avoided the word in the first edition of the *Origin of Species*, preferring to describe biological change as "descent with modification." Taking an uncommon position among nineteenth-century biologists, he did not interpret evolutionary change as inherently progressive.

Thus, the false equation of evolution with progress records a sociocultural bias, not a biological conclusion, and one hardly needs great insight to locate the primary source of this bias in our human desire to view ourselves as the apex of life's history, ruling the earth by right and biological necessity. This fundamental misconception of evolution is strongly abetted by one of the most pervasive of all canonical icons for any scientific concept—the march or ladder of evolutionary progress.

The standard form of this icon—largely a staple of popular culture in cartooning and advertising, but not absent from professional textbooks and museum exhibits—shows a linear sequence of advancing forms (depicted left to right as we read, though my only Israeli example, a recent Pepsi ad, runs right to left). The sequence is shown either globally, running from an amoeba to a white male in a business suit (thus recording another form of iconographic bias), or more parochially as moving from a stooped ape to an upright human. Such a single sequence is, of course, a parody. Most reasonably well-educated people understand that evolution is not a single advancing line. But the caricature works because it epitomizes, by simplification and exaggeration to be sure, the essence of what many people understand by evolution: in a word, progress.

The march of progress has enjoyed an astonishing variety of uses, primarily in commercial humor. I wonder if any other scientific concept is so well and immediately recognized (though in this case almost perversely misinterpreted) by a canonical icon. Consider just two examples from hundreds in my collection. The first (Figure 1) is a favorite of the computer industry. They want to convey the message that their products have gotten smaller and cheaper, so they show a stooped chimp weighted down by a vacuum-tube computer evolving into a white-male-in-business-suit-with-PowerBook. Regional versions also abound, as in *The New Yorker*'s form of Figure 2 (my example from California shows the evolution of swimming trunks through time).

At Toshiba, we've always believed that a computer that merely sits on a desk is a lower form of a computer. That's why five years ago we dedicated ourselves to making the most useful and therefore, the most powerful personal computers of all:

Computers that you can use whenever and wherever you may need them.

Computers that are there for you when you come up with an idea while you're pacing around the airport.

Computers that are there for you when suddenly the client wants to make his five-hundredth revision.

Computers that are there for you at 3:00 A.M. when the business plan is due at nine.

Because those are the times people can really use the power of computing.

And that's why, in our opinion, any computer that can't work where and when people need it, just isn't advanced enough.

We invite you to learn more about why Toshiba is the leader in the next generation of personal computers.

TOSHIBA
In Touch with Tomorrow

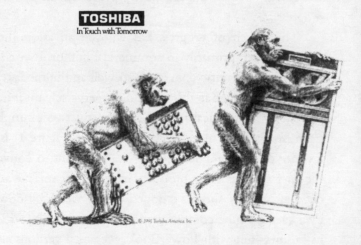

Figure 1 *The march of evolutionary progress*

Figure 2 *Drawing by P. Steiner;* © *1990* The New Yorker
Magazine, Inc.

The power (and recognizability) of the icon is perhaps
shown best by numerous parodies (of the primary parody)
that never fail to be immediately comprehensible. In a Frank
and Ernie cartoon, for example, the standard sequence
runs left to right, from a fish in the sea up a hill to Frank
at the summit, who holds a fishing rod over the cliff to

farther right, and is just about to hook a fish identical with the starting image at extreme left. In my favorite example, an editorial cartoon entitled "Education in the United States," four identical stooped monkeys wearing dunce caps form a single line. Surely an icon has become powerful and canonical when comprehension of a parody depends upon the absence of the original image itself, with the opposite concept encoded into the picture actually shown.

III. High Culture's Version of the Ladder

One might dismiss the pop culture versions as pure misconceptions of a scientifically illiterate mass culture, mistakes that would not be made so readily by well-educated people or by scientifically sophisticated non-professionals. But the closest version we have of evolutionary iconography intended for a more sophisticated culture makes exactly the same errors—more subtly but at the same time even more pervasively. I am thankful to the historian of science Martin J. S. Rudwick for explicitly examining this high-culture genre in his recent and excellent book on iconography of prehistoric life, *Scenes from Deep Time* (University of Chicago Press, 1992). Rudwick ends his survey with an account of this genre's establishment in the nineteenth century; I have extended the analysis to our own time.

High culture's version comprises series of paintings for the history of life in geologically sequential order, one for the Cambrian, one for the Ordovician, etc. In other words, we are not viewing single scenes of a selected moment in prehistoric life, but representations of life's history expressed as a series in proper geological order. The demand for such paintings has been small—primarily in museum murals and coffee-table books, though a modest art market for paintings of prehistoric life has emerged for the first time in our generation.

Moreover, this genre could only have originated in the mid-nineteenth century for two reasons. First, no adequate reconstructions of fossil vertebrates existed before Cuvier's seminal work of 1812. Second, the geological time scale was not well worked out until the 1840s or 1850s. As a result of this limited market and restricted time, the high-culture iconography of sequential painting for life's history is small and manageable. One need not take a sample from a large statistical universe; one can actually survey all major examples for common characteristics and differences.

As a primary conclusion to be drawn from a survey of all influential series, we find no essential variation at all. The same misconceptions are encoded in eerily common ways into all examples—a stunning case for the power of canonical iconography to maintain narrowly prejudicial notions about a subject. Rudwick shows that the first influential series of lithographs was produced by Edouard Riou (1833–1900), also Jules Verne's lithographer, for a famous book on the history of life by the French popularizer, Louis Figuier (1819–1894)—*La Terre avant le déluge*, first published in 1863. (An earlier version was published in 1851 by the German paleobotanist F. X. Unger as *Die Umwelt in ihren verschiedenen Bildungsperioden*; but Unger's work appeared in a very small and expensive edition and his concentration on plants, with very few animals represented, ran counter to another of our parochial prejudices, and therefore limited interest in his work.)

Until the current generation, twentieth-century portrayals of the history of life were dominated by the great American artist-naturalist Charles R. Knight (1874–1953), who virtually owned the genre from the 1920s until his death. (Knight did almost all the major murals in American institutions—New York's American Museum of Natural History, Chicago's Field Museum, and Los Angeles's tar pits museum, for example.) Then, in the 1950s, a Czech duo of artist Zdenik Burian and paleontologist Joseph Augusta published a series of wonderful folio books filled with paintings in color—the first real rival to Knight's hegemony.

The domination of this iconographic tradition by the fallacious theme of progress is even more striking than in the familiar ladders of pop culture imagery—both because the particular pictures, without exception, show the same sequence (leading, at least passively, to the notion that such scenes represent the history of life, rather than one pathway among hundreds of potential and undepicted alternatives), and because greater subtlety of presentation masks the iconographic bias. Both the bias and its invariance can be illustrated by comparing Figuier's original series of 1863 with the most prominent of twentieth-century examples, Charles R. Knight's series, painted for *National Geographic* magazine in 1942 and entitled *Parade of Life Through the Ages*.

52

Figure 3 *Figuier's 'Ideal View of the Earth during the Devonian Period,' from* Earth Before the Deluge *(1863)*

The bias of progress has led all these artists to paint the history of life as a progressive sequence leading from marine invertebrate to *Homo sapiens*. Diversification and stability, the two principal themes of natural history, are entirely suppressed, and the tiny, parochial pathway leading to humans stands as a surrogate for the entire history of life. (One might object less if these artists explicitly stated an intent to show the particular excursion through the evolutionary bush that led to human beings—for then we could only accuse them of parochialism. But a look at the titles, and a reading of their text, clearly shows that they claim to be painting *the* history of life. Figuier's work is called *Earth Before the Deluge*, while Knight's bears the title *Parade of Life Through the Ages*.)

The world of invertebrates occupies the first long stretch of life's geological history, but, in an initial display of pervasive prejudice, invertebrates receive only two or three plates (out of thirty to sixty in total)—see Figure 3 for Figuier's version and Figure 4 for Knight's. Figuier's plate shows another interesting iconographic bias, or rather tradition in this case, by depicting invertebrates as thrown up and drying out on the shore, rather than *in situ* as we might view them in an aquarium.

This inadequate conception had long been traditional in Western iconography and did not yield to the more satisfactory *in situ* view until the aquarium craze of the 1840s and 1850s made such a perspective sufficiently familiar to all. Even the "obviously objective" can be more a matter of artistic convention than "plain truth."

I would not object so strongly to the scarcity of plates showing "invertebrates only" if subsequent paintings continued to include invertebrates along with newly risen vertebrates. But as soon as fishes evolve, we never see an invertebrate again (except in the background, and then only occasionally). How can such a narrow view be justified? Invertebrates didn't go away just because fishes appeared. Invertebrates didn't stop evolving when the history of fishes began. Four hundred million years of invertebrate history are simply expunged from the conventional representation of life through the ages. This immense span includes most of the multicellular history of animal life, including such fascinating events as the death of some 95 percent of all species in the Permian mass extinction some 225 million years ago.

Figure 4 *Knight, Earliest life—marine invertebrate creatures of 530,000,000 years ago*

Fish fare no better. As soon as terrestrial vertebrates appear, artists never again show a fish. But fishes make up more than half of all vertebrate species today, and most of their evolution occurred after terrestrial vertebrates arose. For example, nearly all modern fishes belong to the Teleosti, or higher bony fishes. But teleosts didn't evolve until well after the origin of amphibians and reptiles. So this most important of all events in the evolution of vertebrates, the source of more than half of all living vertebrate species, goes entirely unrecorded in the canonical iconography. Is this the history of life—or just a disconnected sequence of animals judged "highest" because, in genealogy or complexity, they closely approach humans through time—a prejudiced perspective indeed?

The canonical sequence then continues from early amphibians to dinosaurs, usually depicted in mortal combat (contrast Figuier's lumbering creatures of Figure 5 with Knight's more agile dinosaurs of Figure 6, but note the similarity of pose and activity). A canonical plate from the time of dinosaurs also serves as the rule-proving exception. Although no fishes are shown after terrestrial vertebrates arise, convention permits another marine scene dated during the reign of dinosaurs—though the only animals depicted are marine reptiles (ichthyosaurs, plesiosaurs, and mosasaurs), never fishes. In other words, one may draw members of "highest" groups that return to ancestral environments, but never the ordinary, and supposedly superseded, forms of those realms.

And so the sequence continues on its familiar route from dinosaurs, to mammals, to humans (note the similarity

Figure 5 *Figuier's 'The Iguanodon and the Megalosaur (Lower
Cretaceous Period)' from* Earth Before the Deluge *(1863)*

Figure 6 *Knight, 'King-tyrant Lizard, Most Terrible of the Dinosaurs, Locked in Mortal Combat'*

Figure 7 *Figuier's 'Appearance of Man' from* Earth Before the Deluge *(1863)*

Figure 8 *Knight, 'With Flint-tipped Spear, Stone Ax, and Rocks, Neanderthal Men Repel an Invader'*

in dress and pugnacity of Figuier's early people in Figure 7 and Knight's in Figure 8).

The hegemony of conventional imagery is so complete that the sequence of pictures moves on through its exceptionless order no matter what the stated philosophy of the artist, whether the sincere Christianity of Charles R. Knight:

> Those of us whose minds are imbued with a proper amount of religious conviction will detect in this apparent selection [for increased human intelligence] the intervention and assistance of a power higher than ourselves—a certain definite purpose, divine or otherwise, whose control has shaped our destiny.

Or the supposed materialism of communist Czechoslovakia, as depicted by Augusta and Burian in the 1950s:

> From the very beginning of the history of life on Earth we see how life constantly develops and progresses, how it is constantly being enriched by new, ever higher and more complex forms, how even man, the culmination of all living things on Earth, is tied to it by his life.

When an iconographic tradition persists for a full century in the face of such disparate ideologies expressed in accompanying text, then we truly grasp the power of pictures and the hidebound character of assumptions that go unchallenged because they are unrecognized in icons rather than explicit in texts.

IV. The Cone as a Canonical Icon of Diversity

Darwin correctly noted that evolution presents two fundamental problems with potentially different solutions (and certainly, I might add, with disparate iconographies): anatomical change within lineages (solved by Darwin with the principle of natural selection), and diversification of species, or increase in the number of lineages. Darwin called this second issue the "principle of diversity" and he developed no satisfactory solution until the middle 1850s. (This timing helps us to resolve the old mystery of why Darwin, having formulated the principle of natural selection in 1838, delayed publication for more than twenty years. The reasons are complex, and mostly involve Darwin's fear of exposure for the radical philosophy underlying his evolutionary views. But failure to solve the problem of diversity also disturbed Darwin, for he knew that he did not possess a complete theory of evolution while he only grasped anatomical change through natural selection, but had not yet formulated an adequate explanation for the splitting of a lineage into two daughter populations.)

The problem of diversity is so topologically distinct from the problem of transformation that a different iconography must be employed for basic illustration. Just as the ladder provides a canonical icon for transformation misconstrued as progress, the same error of falsely equating evolution with progress yields a canonical icon for diversification: the cone of increasing diversity. This icon is less familiar to the general public, for it does not appear either as a popular version like the ladder or as a more sophisticated, but still nonprofessional, genre like paintings of prehistoric life. Thus, the cone of increasing diversity resides largely in textbooks and professional publications for scientists—but it constrains thought no less.

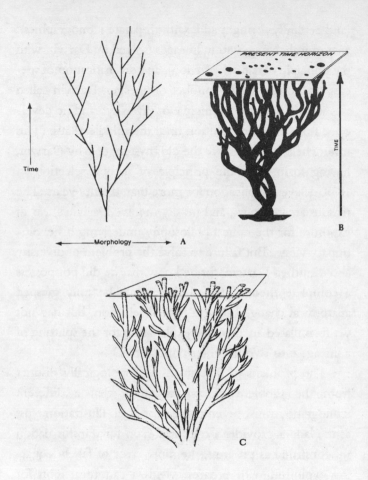

Figure 9 *The iconography of the cone of increasing diversity*

In the cone of increasing diversity, the history of a lineage begins with a single trunk (the common ancestor) and then moves—gradually, smoothly, and continually—upward and outward, occupying more and more space as the number of branches (species) grows (Figure 9 shows typical examples from a modern textbook). But why should such an icon be called biased? What alternative could be suggested? Evolutionary theory demands a common ancestor for related forms, so the tree must emerge from a single trunk at its base. (I accept this argument and regard the common trunk as required by theory, not imposed as a sociocultural bias.)

The biases rather emerge from the canonical shape of such trees above their common trunk—and thus I refer to the canonical icon as a "cone" of diversity. Nothing in theory requires a smooth upward and outward flow for the tree, the feature that sets the tree's shape as an inverted cone or funnel. This arbitrary cone owes its canonical form to several subtle effects of progressivist bias as applied to diversity (rather than to anatomy as in the ladder). First of all, the cone shape requires that the early history of a lineage be composed of only a few major branches, and these must then represent primitive precursors of later forms, thus implying a predictable expansion from limited initial diversity.

Second, and more pervasively, the bias in this canonical icon rests upon a conflation in the meaning of axes. The horizontal axis represents morphology, and greater spread of the tree therefore records expansion in number of species and their adaptations. The vertical axis is supposed to record time alone, so higher branches on the tree

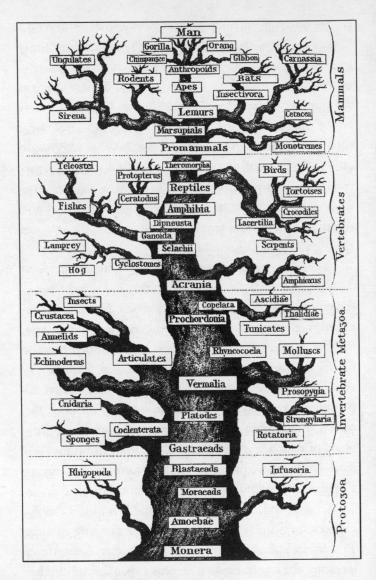

Figure 10 *Haeckel's evolutionary tree*

should represent greater geological youth. But, with the ladder almost inevitably in mind, a higher position on the tree easily becomes conflated with anatomical progress— and the cone of diversity then folds back into the ladder of progress, and the two icons overlap in meaning.

65

If anyone doubts that the cone is a biased icon, consider the first historically important tree of life (bark and all) ever published—Ernst Haeckel's version of 1866 (Figure 10). Haeckel conflates time with progress on the vertical axis, and his tree founders on the logical and pictorial impossibility of adequate representation, at least so long as the cone's dictates are obeyed and the top layer of the tree must therefore spread widest. The bias of progress requires that you place your "highest" creatures in the top layer because you view this lofty place as indicating maximal advance. The cone dictates that this level must bear the most branches. But suppose that the "highest" group is not diverse and contains only a few species. How can you spread them so thin?

Haeckel encounters this insoluble dilemma because he takes a conventional view and regards mammals as superior beings—so he grants them exclusive residence in the top layer of the tree. But mammals are a small group of only 4,000 species or so, and Haeckel, to fill the space, must make fine distinctions, with full branches (and numerous sub-twigs) for whales, carnivores, and, inevitably in the center, primates. But insects, representing almost a million described species, must all occupy a single unbranched twig (more than halfway down at the left) because, as "primitive" forms, they have to be fitted into

66

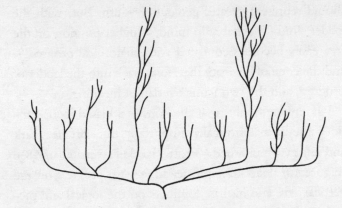

Figure 11 *The revised model of diversification and decimation, suggested by the proper reconstruction of the Burgess Shale*

a lower level of the tree (with much less room on the cone) and must, moreover, share this limited space with other lesser creatures!

Alternatives to such misleading images exist, but the unconscious hegemony of canonical iconography has generally prevented their consideration and the canonical icons have therefore continued to constrain our thinking, for pictures are such powerful guides to our theorizing. (Unconscious hegemony may sound oxymoronic, but such quiet and unobtrusive rule can be the most powerful of all. We all know, after all, that the administration of our offices is most effective when smooth operation remains unnoticed.) For example, in Figure 11, I have tried to draw a non-cone to encompass the very different view of life presented by the full effect of the Cambrian explosion as recorded in the Burgess Shale (see my book *Wonderful Life*, 1989). Here, maximal diversity occurs right near the geological beginning, and life's subsequent history features the loss of most initial anatomical experiments, with concentration of later diversity upon a few surviving designs.

But even this icon of a grass field with most stems mowed and just a few flowering profusely, while circumventing (and almost inverting) the canonical cone, does not capture the most philosophically radical concept arising from our modern study of life's early multicellular history—the notion that most losses occurred by the luck of the draw rather than by the predictable superiority of a few founding lineages, and that any particular lineage still alive today (including our own) owes its existence to the contingency of good fortune. All our canonical icons are based upon the opposite notion of progress and predictability, and therefore preclude proper consideration of contingency as the major force affecting the directions of life.

If icons are central to our thought, not peripheral frills, then the issue of alternative representation becomes fundamental to the history of changing ideas in science (and even to the quite legitimate notion of scientific progress!). How shall we draw the geometry of contingency? How else may we draw the history of life, so that we may come closer to meeting our ancestors face to face and may even probe pictorially into our own psyches to release the potential thoughts that lie even too deep for tears?

Daniel J. Kevles

Pursuing the Unpopular: A History of Courage, Viruses, and Cancer

Students of cancer have been struggling to understand the disease since ancient times. Indeed, cancer derives its name from the Greek for "crab," *karkinos*, expressing the tendency of cancers to claw in multiple directions into normal tissue. Similarly, the study of it—oncology—derives from the Greek *oncos*, for "mass." The physician Galen, whose authority in the matter prevailed at least until 1500 A.D., attributed cancer to an excess of black bile, one of the four humors. Some of his successors found the origins of cancer in—variously—immoral behavior, venery, depression, or (in the case of nuns) celibacy. Others, noting the tendency of some cancers to cluster in families, theorized that cancer was a hereditary affliction. Here and there, from the late eighteenth century onward, several observers suspected that a cause lay in environmental poisons—the soot in which chimney sweepers worked, the snuff and tobacco that gentlemen inhaled, the dust in mines, and the chemicals in aniline

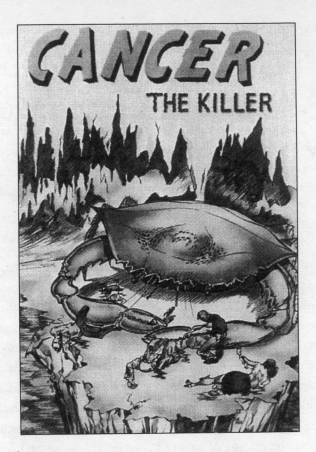

Cancer as a crab

dyes. But at the end of the nineteenth century an honest reporter might have echoed what the leading Philadelphia surgeon Samuel Gross had written about cancer in the middle of it: "All we know, with any degree of certainty, is that we know nothing."[1]

Now, a century later, as a result of a series of incremental steps, we can say that we know something about cancer. No one of these steps suddenly explained the causes of cancer; what each did was to enlarge scientific purchase on understanding of the disease by enabling researchers to pose new questions. As a result, biomedical scientists have begun to understand its mechanisms at the most fundamental level of the living cell. How scientists have come to comprehend—to the extent they do—what makes normal cells into cancerous ones is a strange story of difficulties, disappointments, and frustrations. However, its dramatis personae include one persistent and, as it has turned out, essential class of actors—the viruses that nowadays are called tumor viruses.

By the opening of this century, the prospects for understanding cancer, not to mention numerous other diseases, appeared to brighten with the emergence of scientific medicine, particularly the increasing confirmation, through laboratory experiments, of the theory that specific microorganismic agents are responsible for infectious diseases. To be sure, there were good reasons to think that cancer is not an infectious disease, since neither tumors nor leukemias are transmitted from patient to doctor or nurse or family member. But some scientists held that cancer might involve an infectious agent after observations were made in the late nineteenth century that when tumors taken from one animal were transplanted to another they could provoke new tumor growths.

In the late nineteenth and early twentieth centuries, researchers in many laboratories tried to find cancer-causing microorganisms, searching among protozoa, bacteria, spirochetes, and molds. They all failed, and theories of cancer as an infection were waning in scientific respectability when, in 1909, a farmer brought a cancerous chicken to Peyton Rous, a young biologist on the staff of the Rockefeller Institute for Medical Research. The Rockefeller Institute had been established, in 1901, in New York City to foster research in scientific medicine. Rous, a physician by training, had joined the Rockefeller to pursue cancer research despite the advice of friends who told him it was a hopeless field. The cancerous chicken was a pure-bred barred Plymouth Rock hen,

Peyton Rous

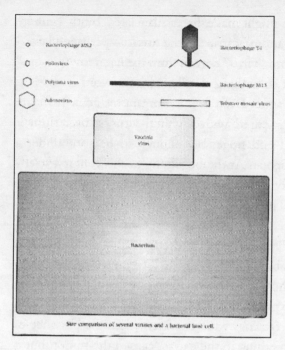

Size comparison of several viruses

with a large irregularly globular tumor protruding from its right breast. The cells were characteristic of a sarcoma, a tumor arising in connective or muscle tissue.

Rous wanted to obtain an extract from the tumor from which both cellular material and bacteria would be eliminated, thereby revealing any agent that might be at work apart from the cells. He did so by mincing the tissue and then filtering it. He then injected a solution of the extract into healthy chickens of the same breed and found that they, too, developed sarcomas. Rous contended that the tumors might have been stimulated by a "minute parasitic organism" carried in the extract—perhaps a virus.[2]

The word "virus" derives from the Latin for "poison." It connotes something deadly, but the word in itself is otherwise vague. Even as microorganisms go, viruses are minute. Compared with a typical virus, a bacterium is complicated and huge, big enough to be seen under a light microscope. At the beginning of this century, scientists could only characterize viruses largely by what they were not. They could not be seen under a light microscope. They could not be isolated from liquids even with the finest filters available. Indeed, they were often called "non-filterable agents"—agents, of course, of infectious diseases, whose existence made it possible for scientists to detect their presence.

Rous's findings were consistent with the presence of a virus. Something in the extract caused the tumor, but it could not be seen and could not be filtered out. The evidence was persuasive enough to set scientists in a number of laboratories on both sides of the Atlantic to searching for viral causes of tumors in other animals—mice, rats, rabbits, dogs—by obtaining tumor extracts and inoculating them into healthy animals. For the most part, the experiments failed; injection with tumor extracts did not produce malignancies except in chickens and several other types of fowl. No one could say why.

Still, the idea that cancer is caused by an infectious microorganism continued to tantalize some scientists. In Denmark in the several years before World War I, Johannes Andreas Grib Fibiger induced stomach cancer

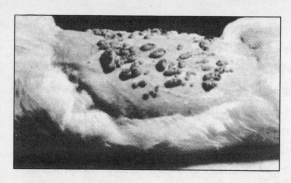

Shope papillomas

in rats by feeding them cockroaches infested with the larvae of the nematode, a tiny parasitical worm. He convinced himself that the nematode, when carried by the cockroach, was an agent of carcinogenesis. Although he convinced few others—not least because hardly anyone could replicate his experiments—he was awarded the Nobel Prize in physiology or medicine in 1926. Fibiger died just two months after he won the prize, and further investigation of his experimental results died with him.[3]

Interest in the role of viruses in causing cancer remained alive at the Rockefeller Institute, where in the early 1930s a scientist named Richard Shope extracted a non-filterable agent from a papilloma—that is, a wart-like skin growth—on a wild rabbit and injected it into domestic rabbits. The domestic rabbits developed papillomas that were at first benign but then became malignant. Researchers, however, regarded such results as anomalies. It was mainly in chickens that viral tumor transmission

experiments had succeeded, and chicken cancers were
believed to have nothing to do with cancer in mammals,
let alone with cancers in human beings. By the 1930s, the
theory that at least some cancers were caused by viruses
had fallen into deep disrepute among most scientific students of cancer, and those who held to the theory risked
their scientific reputations.

This state of affairs imposed severe constraints on
anyone bold enough to pursue research into a possible
relationship between viruses and cancer. Just how severe
was made clear at the Jackson Laboratory, which is lo-
cated on Mt. Desert Island, in Bar Harbor, Maine. Today,
the Jackson Laboratory is the largest non-profit supplier
in the United States of standardized mice for biomedical

Cancer poster

NEW CANCER POSTER

The National Safety Council, cooperating through its Health Service Section, with the American Society
for the Control of Cancer has just published a poster on cancer, of which the text is given below. Additional
copies of this poster may be obtained at cost from the National Safety Council, Chicago, Ill.

About Cancer

One Out of Every Ten Persons Over Forty Dies of Cancer

Cancer Is Curable If Treated Early

Cancer begins as a local disease. *If recognized in time it can often be completely removed
and the patient cured.* If neglected, it spreads
through the body with fatal results. No medicine will cure cancer. Early diagnosis is all important, but pain rarely gives the first warning.

Danger Signals—

(1) Any lump, especially in the breast.
(2) Any irregular bleeding or discharge.
(3) Any sore that does not heal, particularly about the mouth, lip or tongue.
(4) Persistent indigestion with loss of
weight.

These signs do not necessarily mean cancer,
but any one of them should take you to a com-
petent doctor for a thorough examination. Don't
wait until you are sure it is cancer. It may then
be too late.

C. C. Little

research. When it was set up in 1929 it was intended primarily to use mice for cancer research. Its founder was Clarence C. Little, a Boston patrician descended from Paul Revere. Handsome, charming, and charismatic, Little was an accomplished scientist, a former university president, and the half-time managing director of the American Society for the Control of Cancer, the predecessor to the American Cancer Society. At the time, cancer was overtaking tuberculosis as a cause of death, and had become the second major killer in the United States, just behind heart disease. Yet cancer struck so randomly and its origins seemed so mysterious that people who contracted it—especially women with breast or uterine cancer—feared stigmatization. Public discussion of the disease tended to be shrouded in embarrassment, ignorance, and fear. From his platform as director of the cancer society, Little mounted a major campaign of public eduation about the disease, making the cover of *Time*, in 1937, for encouraging open discussion on the subject and,

Mouse with breast cancer

in particular, for urging women to be on the alert for the signs of breast and uterine cancer. Little emphasized that cancer was not a disgrace to be suffered in silence by its victims but a puzzle to be solved by doctors and researchers.[4] (In 1937, Little helped create the federal National Cancer Institute, which has outspent every other agency of disease research in the United States and currently has a budget of roughly $2 billion per year.)

The research program at the Jackson Laboratory emphasized the role in cancer of heredity, an approach that perhaps expressed Little's advocacy of eugenics but that he had been committed to since his days as a young scientist, when he had begun exploring why certain strains of mice display a strong tendency to develop tumors. After the rediscovery, in 1900, of Gregor Mendel's classic work on inheritance, heredity as a factor in shaping organisms could be traced through analysis of genes. Using genetics, Jackson scientists developed what were in effect living research tools—strains of mice that were highly inbred and exhibited a high incidence of cancers of various types, including breast tumors and leukemias. The premise of the Jackson program was that, so far as the occurrence of cancer was concerned, genes counted and viruses did not.

However, a Jackson scientist named John Bittner came upon evidence suggesting that viruses do contribute to some mouse breast cancers. Scientists at the laboratory customarily bred pure strains of mice differing from one another in their frequency of cancer, hoping to find a clue to oncogenesis—the causes of cancer—through hybrid breeding. For example, they crossed mice with a high tendency to contract breast cancer with mice with a low tendency to come down with the disease. The crossing program produced a startling result: if the mother belonged to the high-incidence strain and the father to the low-incidence strain, the offspring would contract breast cancer at a relatively high rate; but if the parental categories were reversed, the incidence of cancer among the offspring would be low. In 1936, Bittner traced the phenomenon to the transmission of a cancer-causing agent in the milk that was transmitted through the mother's

John J. Bittner

suckling of the infant mice. The agent appeared to increase risk of breast cancer in the mice but did not cause it uniformly. While 90 percent of the maternal strain of mice contracted breast cancer, no more than 30 percent of the offspring did. To Bittner's mind, a pre-existing susceptibility to cancer—arising perhaps from hormones —made some of the mice more vulnerable than the others to the cancer-causing agent.[5]

Whatever made them vulnerable, Bittner became convinced that the agent itself was a virus, a conviction that was eventually recognized as correct. However, in the 1930s, he called the agent involved a "milk factor" instead of a "virus." Bittner was hesitant to challenge the genetic model of oncogenesis that dominated thinking at the Jackson Laboratory (a model reinforced by Little's worry that a viral hypothesis would give rise to fear that cancer is contagious, especially through breast feeding). He was also reluctant to defy the prevailing orthodoxy in the scientific community at large that viruses had nothing to do with cancer. Shortly after the founding of the National Cancer Institute, an advisory committee to the Surgeon General concluded that viruses as well as other microorganisms could be disregarded as causes of the disease.[6]

By the early 1940s, scientists in several other laboratories were pursuing research on mammary tumors in mice. The evidence they obtained increasingly suggested that Bittner's "factor" might be a virus. Still, scientists in these laboratories were also reluctant to embrace the viral theory. Like Bittner, they resorted to euphemisms, referring to the apparent carcinogenic agent as a "milk

Ludwik Gross

influence." Perhaps they shared the apprehensions of Bittner, who later noted, "If I had called it a virus, my grant applications would automatically have been put into the category of 'unrespectable proposals.' As long as I used the term 'factor,' it was respectable genetics." Bittner himself began openly speculating that his "factor" was a virus only after he left the Jackson Laboratory in 1942 to join the faculty of the University of Minnesota Medical School. It was not until the late 1940s that a viral interpretation of the milk influence achieved general respectability as a point of serious scientific discussion.[7] By the late 1950s, after considerable debate over complicated and contradictory evidence, it was widely recognized that the "influence" is in fact a virus—the Mouse Mammary Tumor Virus, as it came to be called.

*   *   *

Before then, an outsider to the biomedical research establishment had revived the belief that viruses had something to do with cancer. He was Ludwik Gross, who had arrived in the United States in 1940 as a young Jewish refugee, having escaped from his native Poland upon the outbreak of World War II by driving an automobile ahead of Hitler's advancing armies. Gross had spent much of the Thirties at the Pasteur Institute in Paris, investigating whether leukemia was caused by a virus. Although he found no evidence that this was so, within a few years after he arrived in America he became convinced to the point of obsession that viruses must be responsible for cancers. Commissioned as a medical officer in the US Army and posted to an Army hospital in Tennessee, he planned a new series of experiments that would employ mice and that, he hoped, would prove the point. He obtained the nucleus of an experimental mouse colony from John Bittner, storing the animals in coffee cans and carrying them around in his car when necessary, impatiently waiting for an opportunity to begin the new line of research. Early in 1944, he was assigned to the Veterans Hospital in the Bronx. There, while working with cancer patients, he set up a small laboratory space for himself.

For his experimental plan, Gross proposed to remove organs from leukemic mice, grind them up, and filter them as Rous had done, and inject the extract of ground-up cells into non-leukemic mice. The cells would contain the virus that he presumed caused leukemia. He persisted for four years, working alone in his makeshift laboratory, without grants or any other kind of support—and without success. He was about to abandon the project when one day, probably in 1949, he heard a report that a certain virus would produce paralysis in mice if it was injected into them when they were only forty-eight hours old. Gross had been inoculating adult mice. He says that he knew immediately on hearing the lecture that if he inoculated newly born mice with ground-up leukemic cells, the mice would develop leukemia; their immune systems, he reasoned, would not yet have developed sufficiently to resist the virus. He promptly inoculated a group of infant mice with unfiltered leukemic cells; within two weeks, all the mice developed the disease. In a follow-up experiment, Gross injected healthy newborn mice with filtered extract—that is, ground-up extract from which the cellular material had been filtered out, leaving only the non-filterable agent, the virus. Those mice developed leukemia, too.

Gross published his results in 1951 and 1952. But although the evidence seemed indisputable that he had demonstrated viral transmission of mouse leukemia, most cancer researchers did not take his work seriously. Indeed, some considered him dishonest. Gross has recalled:

> A few [oncologists] even doubted my integrity; one of the well-known pathologists then employed at Memorial Hospital in New York refused to shake my hand when I greeted him before one of my lectures. I was severely, sometimes even viciously, criticized. In the discussions that followed my presentations, and also in separate communications, the leading experts on mouse leukemia stated emphatically that my observations could not be confirmed.[8]

In today's climate, Ludwik Gross might have been charged with scientific misconduct or fraud and called

Jacob Furth

before the Office of Research Integrity, in Washington. But the difficulties that other scientists experienced in reproducing Gross's results did not reflect any lack of probity on his part. His results derived from a combination of persistence, insight, and, in retrospect, astonishing luck. It later turned out that the mouse strain into which he had injected the leukemic virus possessed a genetic makeup that made it vulnerable to the virus. He would have failed with any other strain. George Klein, a Swedish scientist who has written on Gross's work, commented that he "was indeed honest, but because of the general lack of confidence in him, no one bothered to repeat his experiments exactly—to make filtrates from the same kind of leukemia, to use absolutely newborn mice, or to infect the only susceptible recipient strain."[9]

The situation was changed by Jacob Furth, a respected scientist at Cornell University, who had provided Gross with one of the strains of mice that he used. In the mid-1950s, he bothered to repeat Gross's experiments exactly, fully confirmed his results, and notified the world of the fact. Furth's authority convinced biomedical scientists that viruses do indeed have something to do with animal cancers. Once convinced, scientists in many places obtained filtrates from a large variety of tumors, injected them into newborn mice, and isolated an abundance of viruses that provoked tumors in many species, including hamsters, rats, apes, and cats. Because of the invention of the electron microscope, which had become available for biological research in the 1940s, the viruses could now actually be seen in tumor tissue. By the early 1960s, research on animal tumor viruses was flourishing, enlarging the texts published about them and forming a major branch of basic medical and biological science. The non-filterable viral agent that had, in a sense, started it all in 1911 was now called the Rous sarcoma virus, and in 1966, at the age of eighty-five, Peyton Rous shared the Nobel Prize in physiology or medicine.

*　　*　　*

Nobel Prizes are often awarded for achievements in science that are not only meritorious in and of themselves but also open up new avenues of fundamental research. The recognition of Rous was a case in point. One advantage of tumor viruses is that they can be used as instruments of experiment to pry open the mystery of what happens in a living normal cell when it changes into an abnormal one.

The discovery of the structure of DNA, in 1953, and the working out of the genetic code in the following

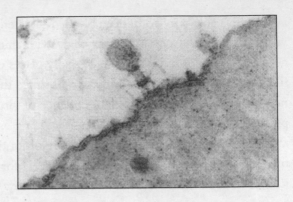

A virus called a bacteriophage injecting its DNA into the bacterial cell (an electromicrograph)

decade had laid the foundation for inquiry into virology at a detailed molecular level. DNA—deoxyribonucleic acid—is, of course, the famed molecule that is built as a double helix. It resides in the nucleus of the cell and carries the genetic information of almost every form of life. The two outer strands of the helix are joined at periodic intervals by rungs fashioned of one of two pairs of bases: adenine (A) and thymine (T) or cytosine (C) and guanine (G). A always pairs with T and C always with G. Each base is thus said to be "complementary" to its pairing counterpart. The four bases form the alphabet of the genetic code. Physically, a gene is a discrete sequence of base pairs along a length of DNA; its genetic information is determined by the ordering of the letters in the code. The DNA of a human being is carried on twenty-three pairs of the rod-shaped bodies called chromosomes, each of which contains thousands of individual genes.

The genetic information becomes operative in a cell through the agency of a molecule called RNA—ribonucleic acid. RNA is a single-stranded molecule, but it resembles DNA chemically, containing bases that are complementary to the bases in a stretch of DNA. A gene's information is transferred from DNA by the formation of a strand of RNA that is complementary to the DNA's coding bases. Acting as a messenger, the RNA bears the information to a part of the cellular machinery that generates a protein determined by the coded instructions.

By the 1960s, a good deal more was known about animal viruses than in the days of Peyton Rous's experiment, enough to begin investigating the mystery of cellular transformation in some detail. Viruses had been demonstrated to be very simple organisms, consisting of either DNA or RNA wrapped in a protein coat. A virus that consists of DNA and causes a tumor is called a DNA tumor virus. Like DNA, RNA can be the source of a virus's primary genetic information; an appropriate sequence of bases along a length of RNA makes up a viral gene. Viruses are so simple that they cannot reproduce by themselves; in order to multiply they need to invade the cell of a higher organism and exploit its generative machinery. Their genetic simplicity and exploitative propensities prompted a number of scientists to pursue the key question of how viruses manage to transform normal cells into cancerous cells, i.e., make them tumorous.

One of the important pioneers among them was Renato Dulbecco, who had trained as a physician in

Howard Temin,
right;
David Baltimore,
opposite page, left;
Renato Dulbecco,
opposite page, right

his native Italy before World War II and came to the
United States in 1947 to work on bacterial viruses. During the 1950s, at the California Institute of Technology,
Dulbecco adapted the concepts and techniques of his
work on bacterial viruses to investigations of animal
viruses such as the polio virus and then to tumor viruses.
In 1960, in collaboration with Marguerite Vogt, a research
fellow at Caltech, he produced tumors in hamster cells by
growing them in a culture with the polyoma virus, which
had been discovered in 1957 and whose name expressed
its ability to transform cells in several different organisms.
Dulbecco and Vogt observed that the polyoma virus quit
reproducing once it started to transform the hamster cells.
Dulbecco hypothesized that the genetic material of the
polyoma DNA was no longer available for viral reproduction because, in transforming the cell, it incorporates itself
into the cell's native DNA and—here was the mechanism
of viral transformation—perverts the cell's machinery of
regulated growth so that it multiplies malignantly. The
hypothesis was plausible in principle, partly because

bacterial viruses were known to integrate themselves into bacterial DNA. In 1962, Dulbecco left Caltech for the new Salk Institute in La Jolla, California, where, during the next several years, he and his collaborators proved that what could happen in principle happened in fact.[10]

Howard Temin, a researcher at the University of Wisconsin, had a similar idea about how the Rous sarcoma virus transforms chicken cells, but his theory presented many problems because the genetic core of the Rous virus is RNA, not DNA. It was plausible that viral DNA could integrate into the DNA of a cell, but such integration was thought to be physically impossible for RNA from any organism. Nor was RNA held to be capable of generating DNA complementary to itself that could be integrated into the DNA of a cell. According to the central dogma of molecular biology at the time, while DNA could generate RNA complementary to it, the reverse did not occur. Genetic common sense declared that without integration into the DNA of a cell there could be no hereditary transformation. Nevertheless, the genetic material of the Rous tumor virus was clearly RNA and it did permanently transform the line of cells deriving from the original invaded cell. In 1962, at a symposium on the biology of animal viruses, Dulbecco remarked that to deal with RNA viruses, scientists had "to blaze new trails most of the time."[11]

Temin, who was willing to go wherever his experimental evidence led him, wrote in an article about the Rous virus in 1964, "The virus acts as a carcinogenic agent by adding some new genetic information to the cell."[12] In Madison several years ago, he told me that "intellectually I felt that the central dogma was true, but that it didn't explain my results. Since this is biology, I didn't have any philosophical problems with my results being an exception—biology doesn't have the force of physics." While a junior faculty member at Wisconsin, Temin openly advanced the unorthodox hypothesis that the Rous viral RNA did, in fact, generate DNA complementary to itself that could integrate into the DNA of a cell. Furthermore, the new DNA constituted what Temin called a "provirus," a piece of DNA coded to produce a daughter virus, including its protein coat and RNA core.[13] The hypothesis that a provirus could be created in this way was perhaps Temin's most striking new conception. The sequence of the virus's reproduction then seemed clear to Temin: on entering a cell, the virus somehow produces a strand of DNA that is complementary to the cell's RNA; the new DNA forms a provirus that can in turn produce a new virus by enlisting the machinery of the infected cell.

Temin's view was widely held to be scientifically bizarre and wrongheaded. He published experimental results indicating that proviral DNA integrates into the DNA of an infected cell, but his data were regarded as weak. No one said so to his face, but he told me how, for example, one prominent virologist, reviewing work on the Rous sarcoma virus at a meeting, said, "I'll give Howard's idea the amount of time it's worth—none." Such ridicule did not demoralize him or interfere with his ability to obtain grants, he told me, but it did shape his research. It led him between 1964 and 1969 to intensify his efforts to prove that a DNA complement to the Rous viral RNA does become integrated into the host cell's genome. He might instead have attempted to explain how the RNA of a virus such as the one in Rous's sarcoma could be made into DNA, but he was not attentive to the particle—the virus—as such, let alone to its capacity to generate the RNA–DNA switch. This was, he acknowledged, a mistake. "I wish I could tell you that I was cleverer. I wish I could tell you that I was smarter than I was," he said to me.

In 1969, Temin shifted his attention to the virus itself, particularly to the question of how it generated DNA from its RNA. Once he turned to this question, he rapidly found the answer: The virus contains a special type of protein called an enzyme that catalyzes the synthesis of DNA from RNA.

The same discovery was made simultaneously and independently by David Baltimore, a member of the MIT faculty who had spent almost three years as a research associate at the Salk Institute with Dulbecco. Baltimore's interest in biology dated from a high-school summer at the Jackson Laboratory, during which he met Temin, who was four years his senior. Since his graduate school days at MIT, he had been working on the genetic systems of RNA viruses—first the polio virus and then, in 1969, a virus called VSV—concentrating on the molecular dynamics of their reproduction. In 1970, in collaboration with Alice Huang, his wife, he discovered that VSV contains an enzyme that permits it to replicate by catalyzing the creation of an intermediate form of RNA that is complementary to its own; the intermediate RNA then generates the RNA of a new virus. That revelation stimulated him to wonder if such an enzyme might also account for how RNA tumor viruses multiply. He was well aware of Temin's theory that RNA tumor viruses reproduce by creating an intermediate "provirus" in the DNA of the host cell. Baltimore later recalled that although Temin's logic "was persuasive, and seems in retrospect to have been flawless, in 1970 there were few advocates and many suspects," adding, "Luckily, I had no experience in the field and so no axe to grind—I also had enormous respect for Howard dating back to when he had been the guru of the summer school I attended at the Jackson Laboratory." Hedging his bet, Baltimore looked for an enzyme in RNA tumor viruses that would make either RNA or DNA. He found evidence of an enzyme that could catalyze DNA in the leukemia virus of a mouse; then, although he at first failed to find the same enzyme in the Rous sarcoma virus, he also found it there too.[14]

In 1970, Temin and Baltimore reported, in separate articles published in the same issue of *Nature*, the discovery of the enzyme, which was promptly dubbed "reverse transcriptase" in recognition of its ability to transcribe RNA back into DNA. RNA viruses came to be termed "retroviruses" because they are equipped with the enzyme to accomplish the transcription. The achievement was worthy of a Nobel Prize—Baltimore, Dulbecco, and Temin shared the award in 1975—partly because it cast a brilliant light on the means of retroviral replication, adding convincing force to Temin's hypothesis of a provirus system by which retroviruses reproduce themselves. The discovery was also important because it opened up further rich lines of research in the interactions of viruses and cells, including the types of interaction that produce tumors. Temin noted at the end of his paper announcing reverse transcriptase that the discovery raised "strong implications for theories of viral carcinogenesis."[15] It did not reveal the cause of cancer, but it supplied an additional experimental tool for prying apart its originating mechanisms.

*   *   *

The research program that led to the understanding of RNA tumor viruses had proceeded in basic biological research laboratories that were, for the most part, only loosely connected, if at all, to the clinical research in which medical scientists dealt directly with human cancer. By the beginning of the 1970s, the medical scientists knew more than their turn-of-the-century predecessors about the etiology of cancer. They knew that the disease could be provoked by, for example, asbestos in the workplace, certain chemicals in food additives, cigarette smoking, and radiation. But they remained largely in the dark about the specific mechanisms of such carcinogenesis or about the causes of cancers that appeared to occur spontaneously or by hereditary predisposition. Basic biologists believed that work with tumor viruses would shed light on all these issues eventually, but for most oncologists and their patients they seemed of no help at all.

To many medical researchers, analyses of RNA tumor viruses appeared irrelevant to the immediate problem of human cancer, especially the challenge of finding therapies or cures for the disease. In animals, most tumor viruses produce sarcomas and leukemias, but most human cancers—especially the lethal ones—are carcinomas, the type of killer tumors that show up in, for example, the lung, breast, colon, prostate, rectum, bladder, uterus, pancreas, stomach, cervix, ovary, kidney, and brain. With possibly one exception—the Epstein-Barr virus, which is implicated in Burkitt's lymphoma—no virus had been shown to cause cancer in human beings. The advances in understanding of animal tumor viruses encouraged a number of basic cancer researchers to think that perhaps they might find viral causes of human cancers and to hope that, if they did, vaccines might be developed against them. But most basic virologists believed there were no human retroviruses. Medical researchers argued that even if they did exist and vaccines could be developed against them, the vaccines would do nothing to prevent the many cancers that originated from non-viral agents such as chemicals or that occurred spontaneously.

Such arguments were put forward in 1968, in an influential book titled *Cure for Cancer: A National Goal* by Solomon Garb. A pharmacologist and physician at the University of Missouri Medical School, Garb argued that determining the mechanisms of a disease was not a pre-requisite to dealing with it. He claimed that cancer researchers needed to give far more attention to therapies and cures, including improvements in surgery, radio-therapy, immunotherapy, chemotherapies derived from plant materials and sea organisms, and Vitamin C. Insist-ing that a national commitment to curing cancer was at least as important as sending a man to the moon, Garb warned that "virtually no consideration" had been given to "the possible political impact of a Russian discovery of a cure for cancer."[16]

Garb's book captured the attention of the medical patron and activist Mary Lasker, who in 1969 mounted a campaign for a more aggressive federal cancer program that rapidly earned support in Washington.[17] In his State of the Union address in January 1971 President Nixon enthusiastically compared the initiative to the Manhattan Project and the Apollo Program, and on December 23, 1971, he signed the National Cancer Act. By 1976 the so-called War on Cancer would more than triple the budget of the National Cancer Institute.

In Congressional hearings and public statements, many biomedical scientists had opposed the kind of ambitious

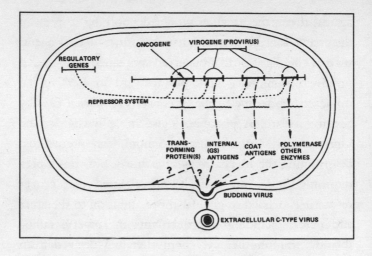

Huebner and Todaro's oncogene hypothesis in graphic form

anti-cancer campaign that Garb wanted, calling it techni-
cally unjustifiable. There had been no biological revelation
—nothing comparable to, say, the discovery of nuclear
fission—to warrant a declaration of war on cancer. To
some biologists, the commitment of vast resources to
such an undertaking was, to say the least, inappropriate
in view of the war in Vietnam and upheavals in the inner
cities. Howard Temin, in a 1972 review of research on
RNA tumor viruses, summarily declared that taking money
from the defense budget to expand cancer research would
constitute "a double benefit to our society." But expansion of the war on cancer should not come "at the expense of the vast domestic social needs that are tearing our
society apart."[18] Although the War on Cancer was enacted anyway, in their dealings with both Congress and

federal agencies, the scientists were powerful enough to keep the War on Cancer program from concentrating exclusively on hit-or-miss objectives such as the ones proposed by Garb.

Part of the newly appropriated money was used to finance research on human cancer viruses. Only a few were found—the number definitively implicated in human cancers remains tiny today—but the processes developed to identify them had important, though unexpected, long-term consequences. By the early 1980s, the laboratory of Robert Gallo, at the National Cancer Institute, had conclusively identified two retroviruses—the RNA viruses HTLV-1 and HTLV-2—as the causes of human leukemia. Gallo has recalled that by 1982 no one doubted the existence of human retroviruses, and within a few years more, few doubted that a retrovirus—HIV—was the cause of AIDS. Gallo has pointed out that "experiences with these earlier human retroviruses gave us the necessary background, knowledge, and credibility to propose a retroviral cause of AIDS and an outline of how to approach the problem."[19]

The War on Cancer also provided considerable funds to promising research in basic tumor virology. To many basic biomedical scientists, cancer could be defeated only through the kind of deeper understanding that might now, in the wake of the discovery of reverse transcriptase, be acquired from studies of the interaction of tumor viruses with the cell. Determining the detailed characteristics and behavior of retroviruses presented new scientific opportunities. What was needed, first, was to confirm Temin's proviral hypothesis by, for example, checking for direct evidence—better evidence than Temin had accumulated —of the presence of proviral DNA in infected cells. Still more tantalizing was the essential question: Exactly how does the integration of the virus's genetic information into the cell's DNA turn it into a source of malignancy?

In 1969, Robert J. Huebner and George J. Todaro had advanced an ingenious hypothesis bearing on this question. Huebner, a distinguished senior virologist who had won a National Medal of Science that year, was chief of the RNA Tumor Virus Laboratory at the National Cancer Institute; Todaro was a junior staff member. Their theory tried to explain how RNA tumor viruses such as the Rous virus have a part in the natural occurrence of tumors. They were struck by revelations obtained with the electron microscope, as well as by other means, about tumors in a number of vertebrate species such as guinea pigs, rats, swine, and snakes. The tumors were not the result of new viral infection but had occurred either spontaneously or as a result of provocation by physical or chemical agents. Nevertheless, they appeared to contain RNA tumor viruses. Huebner and Todaro thus proposed that the cells of many, if not all, vertebrates must naturally contain what they called "virogenes"—DNA that would generate the observed RNA viruses. Furthermore, they contended that the virogenes must include an "onco-gene," a gene capable of inducing the tumors in which the viruses—which themselves had no role in causing the cancer—were found. They speculated that the virogenes had entered the cell and its DNA by some ancient infection and had been hereditarily transmitted from one generation to the next. They suggested that in the normal cell virogenes are repressed but that they might undergo a "de-repression"—and express themselves—as a result of either natural causes or environmental carcinogens.[20] When such expression took place, the result was both the production of the observed viruses and the transformation of normal cells into tumorous ones.

Other scientists had speculated that something in the cell's DNA might be responsible for setting off the chain of events that leads to cancer, but Huebner and Todaro's theory attracted attention, no doubt partly because of Huebner's professional distinction, but also because it provided a genetic model that comprehensively tied together known provocations of cancer—hereditary disposition and environmental carcinogens—along with the viral evidence. Scientists challenged the theory for lack of clarity: What precisely were oncogenes? And how exactly were virogenes de-repressed? Still, whatever its vagueness, Huebner and Todaro's hypothesis enlarged the scope of proviral research to include searches in the cell not only for viral DNA but also for some part of it that might be oncogenic.

The hypothesis intrigued J. Michael Bishop and Harold E. Varmus, who were young faculty members at the University of California, San Francisco, Medical School. Both were physicians who had worked at the National Institutes of Health, where each had become absorbed in the basic molecular biology of viral behavior. Varmus had come to San Francisco in 1970 to work as a postdoctoral fellow with Bishop, who was only three years his senior and had arrived there in 1968. Bishop, the son of a Lutheran minister from small-town Pennsylvania, was quite different in background from Varmus, a doctor's son who grew up outside New York City in a family of second-generation Eastern European Jewish immigrants. But the two shared important interests outside science. Varmus, an anti-establishment editor of his Amherst College newspaper, had done a year's graduate work in Anglo-Saxon and metaphysical poetry. Bishop reads voraciously in literature, philosophy, and history; and he has said that "if offered reincarnation, I would choose the career of a performing musician with exceptional talent, preferably in a string quartet."[21]

As scientists, both combine bold imagination with scrupulous attention to experimental evidence. By 1970, each had independently developed an eagerness to investigate replication of retroviruses, especially with the Rous sarcoma virus. Varmus's first experiments in San Francisco sought to obtain direct evidence that proviral genetic information is contained in the DNA of the cells of chickens, because they are so readily susceptible to infection by the Rous tumor virus. The relationship of the two men, as Bishop said later, "evolved rapidly to one of co-equals"[22]—and into a collaboration that, subsidized in significant part by War on Cancer money, led to a surprising and fundamentally significant new chapter in the story of viral oncogenesis.

Bishop and Varmus's common research interests had quickly led them to conclude that they must test the hypothesis that viral oncogenes existed in some form in the normal chicken cell. In principle, the task was simple. The Rous virus was presumed to contain at least one gene that would transform the cell. You needed only to find and isolate that gene, then check whether any stretch of cellular chicken DNA resembles it, which is to say that the viral gene and the stretch of chicken DNA would have a significant number of base-pair sequences in common. Scientists call two such stretches of DNA or RNA "homologues" of each other. Although not identical, the homologous stretches resemble each other much as do cousins who share several features in common indicating that they are blood relations. However, at the time Bishop and Varmus began their joint research, in the early 1970s, identifying and isolating a gene, even in an organism as simple as a virus, was by no means a straightforward process. Identifying a gene was often a matter of detecting mutations: if an organism showed up that looked or behaved differently from its brethren, chances were that one of its genes had mutated. Such a manifestation of genetic change—an unpredictable event that you had to be lucky and observant enough to notice—would thus identify the gene and make possible its isolation and analysis.

As it happened, Bishop and Varmus were in close touch with a group of retroviral investigators on the West Coast who had their share of good luck and were sharply observant. In 1971, one of them—Peter Vogt, at the University of Southern California—came upon a mutant of the Rous sarcoma virus that was unable to transform cells. Another, Peter Duesberg, at Berkeley, showed that the

RNA in the mutant was about 15 percent shorter than the RNA in the normal, transforming virus. Vogt and Duesberg took the evidence to mean that the missing RNA must contain the unmutated virus's transforming gene, or genes. The evidence was indirect and their interpretation that it indicated one or more genes was an informed guess. The interpretation seemed plausible to Bishop and Varmus; but in order to indicate that their belief was tentative, Varmus told me, they called the missing RNA the "sarc" gene, departing from the standard three-letter nomenclature for a gene used in classical genetics. Whatever it was, having been presented the gift of it by Vogt and Duesberg, Bishop and Varmus proposed in 1972 to determine whether the viral "sarc" gene has a cousin— that is, a homologue—in the DNA of a normal chicken.

Dominique Stehelin, a French postdoctoral fellow working with Bishop and Varmus, used reverse transcriptase to construct a sample of "sarc" DNA from the virus's RNA; he then used a standard procedure to compare the sample with DNA taken from a healthy chicken. The comparison produced decidedly unexpected results. As Bishop, Varmus, and their colleagues reported in 1976, the viral "sarc" fragment is, indeed, homologous to a region in the DNA of chickens. What was surprising was that it is also similar to a strip in the DNA of quail, turkeys, ducks, and even emus, one of the most primitive birds. Unquestionably, "sarc"-related DNA resides in the normal cellular DNA of many avian species. Bishop and Varmus, joined by another postdoctoral fellow, Deborah Spector—Stehelin had gone back to France—went on to check for the presence of the "sarc"-like genes in the cells of mammals and fish. In 1978, they reported that they had found such cousins to "sarc" in the DNA of calves, mice, and salmon. They even had detected evidence of them in human DNA. "Sarc" seemed to have cousins in DNA everywhere.

Bishop and Varmus recognized almost immediately that "sarc" homologues are obviously not the oncogenes —that is, the repressed viral oncogenic DNA—that Huebner and Todaro had originally proposed. As Bishop put it later, they are not "viral genes in disguise," waiting to unmask themselves by transforming a cell.[23] They are only cousins to the viral gene. Their virtual ubiquity led Bishop and Varmus to use evolutionary reasoning to think about what they might in fact be. In evolutionary terms, the birds in which the "sarc" cousins had been found are members of widely divergent species. More striking, according to the fossil record, the major groups of species—birds, mammals, and fish—whose DNA contains the "sarc" cousins had separated at least 400 million years earlier. To Bishop and Varmus, the plain evidence that the "sarc" homologues had been conserved through so much time and speciation indicated that they might be involved in some critical cellular function such as growth and development. They appeared to be not viral genes but normal genes that can be turned into oncogenes.

Bishop and Varmus suggested that the transformation of the cellular cousin of viral "sarc" into an oncogene might be accomplished by the retrovirus through the process that scientists call "transduction." Upon entering the cell, the retrovirus captures the DNA of the normal

Bishop and Varmus

cellular gene, in the process disrupting its ability to function normally. However, Bishop and Varmus also argued, in 1977, that the cellular gene does not necessarily need the virus to become oncogenic. It might also be perverted by a physical or chemical agent.

In 1979, the laboratory of Robert Weinberg, at MIT, reported experimental evidence that normal cellular DNA could indeed be transformed into oncogenic DNA by chemical means. Weinberg and his collaborators accomplished the trick by treating a line of mouse cells with a chemical carcinogen.[24] Soon, experiments in other laboratories detected transformed DNA —that is, DNA changed from that of a normal cellular gene—in a variety of cancer cells, including carcinomas taken from rabbits, rats, mice, and people. For a time, it was thought that two different types of normal cellular genes could be transformed: those that could be made oncogenic by viral action and those that could be made so by other agents, such as chemicals. However, it was clear by 1983 that no such distinction is warranted. A number of cellular genes had been identified that are cousins to both viral oncogenes and to the oncogenes found in cancers, including human cancers, that are not induced by viruses.

Most of these cellular genes seem to exist all over the tree of animal evolution, just like the cousins to the viral "sarc" gene (which is truly a gene and is now spelled *src*, in conformity with the classical rules of three-letter genetic nomenclature). As Bishop and Varmus had suggested for *src*-related genes, these other cellular genes are probably involved in the fundamental cellular processes of growth, regulation, and differentiation. They are thus normal cellular genes that can be turned into oncogenes by chance processes within the cell as well as by environmental carcinogens, and, in a few cases, viruses. They are now termed "proto-oncogenes." They are potential agents of cancer and, as such, Bishop has said, are a kind of "enemy within."[25]

The multiple steps that led to the discovery of oncogenes amounted to a revolution—not only in the study of cancer but in approaches to the processes of normal cellular growth and regulation. In 1989, Bishop and Varmus won the Nobel Prize for their work on cellular genes. Some kind of recognition might also have gone to the Rous sarcoma virus. Although not a cause of human cancer, it has been a decisive key to cancer mechanisms. It was a reliable agent of analysis, a guide through the seventy-year journey, with its lonely detours and divagations, that led from Peyton Rous's identification of his non-filterable tumor-causing extract, through Temin and Baltimore's discovery of the retroviral mechanism, to the discovery of the proto-oncogenes, some 100 of which have now been identified.

It is difficult to think of another case of scientific advance where almost every one of the key pioneers encountered pointed resistance from his community of peers. Unlike earlier episodes in the history of science, the resistance originated in neither religious nor ideological prejudice. It derived from the skepticism of a professional community of biomedical scientists whose beliefs were grounded in available laboratory evidence. The nature of the resistance varied. Rous, Bittner, and Gross had to persuade their peers that animal tumors could be caused by tumor viruses. Temin had to persist against ridicule of his claim that RNA could somehow generate DNA. Huebner and Todaro had to be bold enough to propose a general theory of oncogenesis to basic biologists, who are disinclined to general theorizing and do not hold it in high regard. Even Bishop and Varmus were scoffed at in some quarters for suggesting that their discovery of oncogenes in animal tumors might have something to do with human cancers.

What permitted the pioneers eventually to prevail was to a significant extent their professional courage, imagination, and persistence. Yet it was also the tolerance

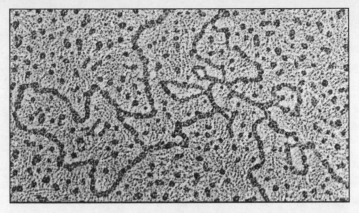

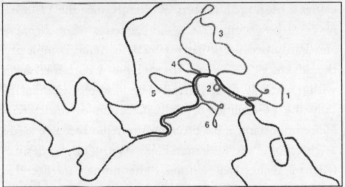

Electron micrograph (top) of oncogenes and accompanying drawing (bottom) distinguishing cellular and viral oncogenes

and pluralism of the basic biomedical research system—the tolerance of deviant ideas and the pluralism that provides niches (large like Rous's and Temin's or small like Gross's) in which the ideas have a chance to flourish. No less important, what led to the illumination of the problem of human cancer was the assumption of the essential unity of life that has been built into the research system.

This assumption, which is central to post-Darwinian biological thought, has been steadily reinforced by the findings in fields such as physiology, biochemistry, and molecular biology. Clearly, human beings are very different from chickens or mice. Still, because their vital processes have so much in common, studies of non-human organisms can shed light on the human one—as they ultimately did in the discovery of oncogenes.

That discovery hardly accomplished a complete understanding of the mechanisms of cancer. No single gene causes most human cancers. Research since the 1970s has revealed how complicated the processes of oncogencsis are. Partly because it has revealed so much complexity, the advent of oncogenes has not put a cure for cancer within reach; but it has already begun to change dramatically the possibilities of early detection of the disease. More important, it has brought biomedical science across a threshold of understanding. With oncogenes, as Michael Bishop has written, "the human intellect has finally laid hold of cancer with a grip that may eventually extract the deadly secrets of the disease."[26]

The next steps toward resolving the mystery of cancer include identifying the specific role of proto-oncogenes in normal cellular processes, determining with greater exactitude what causes the transformation of such genes into oncogenes, and examining why this transformation turns normal cells into tumorous ones. Such knowledge would open the door to much more precise and, hence, effective biochemical interventions against cancer and possibly to genetic therapies for it. But since cancer is a distortion of normal cellular growth and regulation, mastery of the disease will probably come only with a full understanding of the highly complicated processes of normal cellular development. If history is any guide, the extraction of cancer's ultimate secrets, and the development of the therapies and cures that will presumably accompany it, will not likely be achieved by simply throwing money against one or another expression of the disease. Most probably, the mechanisms of human cancer will be exposed by the kind of incremental and often roundabout progress, depending on research in both humans and non-human organisms, that has been made to date—the kind that permits us to say, unlike Samuel Gross more than a century ago, that we know something about cancer, and that we will likely know a great deal more a century hence.

Author's note: I am indebted to several people for critical comments on various drafts or parts of this essay: J. Michael Bishop, Charles Galperin, Jean-Paul Gaudilliere, Michel Morange, Ray Owen, Sondra Schlesinger, and William Summers. I am also grateful to the Andrew W. Mellon Foundation for research support and to Robert Silvers for his editorial work and encouragement.

Notes

1 Quoted in James T. Patterson, *The Dread Disease: Cancer and Modern American Culture* (Harvard University Press, 1987), p. 21.

2 Three years earlier, two Danish investigators had transmitted a leukemia between fowl by injection of cell-free filtrates, but their work was ignored, partly because leukemia was not then thought to be related to cancer. Ludwik Gross, *Oncogenic Viruses* (2nd ed.; Oxford: Pergamon Press, 1970), pp. 99-103; Peyton Rous, "A Sarcoma of the Fowl Transmissible by an Agent Separable from the Tumor Cells," *Journal of Experimental Medicine*, Vol. 13 (1911), pp. 397–403, 408-409.

3 Paul Weindling and Marcia Meldrum, "Johannes Andreas Grib Fibiger, 1926," in *Nobel Laureates in Medicine or Physiology: A Biographical Dictionary*, edited by Daniel M. Fox, Marcia Meldrum, and Ira Rezak (Garland Publishing Co., 1990), pp. 177–181.

4 See Clarence Cook Little, *Civilization Against Cancer* (Farrar & Rinehart, 1939).

5 George Klein, *The Atheist and the Holy City: Encounters and Reflections*, translated by Theodore and Ingrid Friedman (MIT Press, 1990), pp. 120–122.

6 Ibid., p. 122; Kenneth E. Studer and Daryl E. Chubin, *The Cancer Mission: Social Contexts of Biomedical Research* (Sage Library of Social Science, Vol. 103, 1980), p. 21.

7 Klein, *The Atheist and the Holy City*, p. 122; Jean-Paul Gaudilliere, "Cancer between Heredity and Contagion: About the Production of Mice and Viruses," to appear in Everett Mendelsohn, editor, *Human Genetics* (Sociology of Science Yearbook, forthcoming).

8 Marcel Bessis, "How the Mouse Leukemia Virus Was Discovered: A Talk with Ludwik Gross," *Nouvelle Revue Française d'Hématologie*, Vol. 16 (1976), p. 296.

[9] Klein, *The Atheist and the Holy City*, p. 127.

[10] Marguerite Vogt and Renato Dulbecco, "Virus-Cell Interaction with a Tumor-Producing Virus," *Proceedings of the National Academy of Sciences*, Vol. 46 (1960), p. 369; Renato Dulbecco, "Basic Mechanisms in the Biology of Animal Viruses: Concluding Address," *Basic Mechanisms in Animal Virus Biology*, Cold Spring Harbor Symposia on Quantitative Biology, Vol. XXVII (Cold Spring Harbor Biological Laboratory, 1962), p. 523; Renato Dulbecco, *Aventurier du Vivant* (Paris: Plon, 1990), pp. 179–181, 190–192, 208–216.

[11] Dulbecco, "Basic Mechanisms in the Biology of Animal Viruses," p. 520.

[12] Howard M. Temin, "Homology between RNA from Rous Sarcoma Virus and DNA from Rous Sarcoma Virus-Infected Cells," *Proceedings of the National Academy of Sciences*, Vol. 52 (1964), pp. 323–329.

[13] Temin adapted the word "provirus" from "prophage," a term that had been coined to describe a type of phage, which is a virus that preys on bacteria; see Charles Galperin, "Virus, Provirus et Cancer," *Revue d'Histoire des Sciences et de leurs Applications*, Vol. 47 (1994), pp. 7–56.

[14] David Baltimore, "Viruses, Polymerases and Cancer," Nobel Lecture, December 12, 1975, in *Les Prix Nobel en 1975* (Stockholm: The Nobel Foundation, 1976), pp. 159–160.

[15] Baltimore later noted that in the framework of reproduction—the essential context for any organism from an evolutionary point of view—"the ability of some retroviruses to cause cancer is a gratuitous one," but also one that from a human point of view is fundamentally important. Ibid., p. 164; Howard M. Temin and Satoshi Mizutani, "RNA-dependent DNA Polymerase in Virions of Rous Sarcoma Virus," *Nature*, No. 226 (1970), pp. 1211–1213.

[16] Solomon Garb, *Cure for Cancer: A National Goal* (Springer, 1968), p. 15.

[17] Richard A. Rettig, *Cancer Crusade: The Story of the National Cancer Act of 1971* (Princeton University Press, 1977), p. 77.

[18] Temin, "The RNA Tumor Viruses—Background and Foreground," *Proceedings of the National Academy of Sciences*, Vol. 69 (April 1972), p. 1019.

[19] Robert Gallo, *Virus Hunting: AIDS, Cancer, and the Human Retrovirus, A Story of Scientific Discovery* (Basic Books, 1991), p. 203.

[20] Robert J. Huebner and George J. Todaro, "Oncogenes of RNA Tumor Viruses as Determinants of Cancer," *Proceedings of the National Academy of Sciences*, Vol. 64 (September–December 1969), pp. 1087–1094. See also George J. Todaro and Robert J. Huebner, "The Viral Oncogene Hypothesis: New Evidence," *Proceedings of the National Academy of Sciences*, Vol. 69 (April 1972), pp. 1009–1115.

[21] J. Michael Bishop, "Retroviruses and Oncogenes, II," *Les Prix Nobel* (Stockholm: The Nobel Foundation, 1990), p. 219.

[22] Bishop, "Retroviruses and Oncogenes, II," p. 218.

[23] Bishop, "Cellular Oncogenes and Retroviruses," *Annual Reviews of Biochemistry*, Vol. 52 (1983), pp. 320-321.

[24] C. Shih et al., "Passage of Phenotypes of Chemically Transformed Cells via Transfection of DNA and Chromatin," *Proceedings of the National Academy of Sciences*, Vol. 76 (1979), pp. 5714-5718.

[25] Bishop, "Enemies Within," *Cell*, Vol. 23 (January 1981), p. 5; Michel Morange, "The Discovery of Cellular Oncogenes," *History and Philosophy of the Life Sciences*, Vol. 15 (1993), pp. 45–58.

[26] Bishop, "Cancer Genes Come of Age," *Cell*, Vol. 32 (April 1983), p. 1020.

R. C. Lewontin

Genes, Environment, and Organisms

Before the Second World War, and for a short time after it as a consequence of the immense notoriety of the atom bomb project and the promise of nuclear energy, physics and chemistry were the sciences of greatest prestige and the image of what natural science should be. When Americans were polled about the relative prestige of various occupations, they rated chemists and nuclear scientists above all other branches of learning, and even practitioners of such "soft" disciplines as psychology and sociology were rated above mere biologists.[1] The philosophy of science was essentially the philosophy of physics, and in his seminal work on the sociology of science, *Science and the Social Order*, Bernard Barber could write that "biology has not yet achieved a conceptual scheme of very high generality like that of the physical sciences. Therefore it is less adequate as a science."[2]

We have changed all that. It is biology that now fills the science columns of national newspapers, and television's fascination with billions and billions of stars has given way to a concentration on the sex lives of thousands of species of animals. The philosophy of science is now largely a consideration of biological issues, especially those raised by genetics and evolutionary theory. The cleverest science students now choose careers in molecular genetics rather than nuclear physics and it is a fair guess that more people can identify Watson and Crick than they can Bohr and Schrödinger.

In part this new dominance of biology comes from our preoccupation with health, but largely it comes from biology's claim to have become an "adequate science" by fulfilling Barber's demand for a "conceptual scheme of very high generality." At the level of molecules, all life is the same. DNA in its various forms is said to carry the information that determines all aspects of the life of all organisms, from the form of their cells to the form of their desires. The DNA code is "universal" (or nearly so): that is, the same DNA message will be translated into the same protein in every species of living being. At the level of organisms, the apparent profligate variety of shapes and ways of making a living, of nutrition and fornication, are all explained as optimal solutions to problems posed by nature, solutions that maximize the number of genes one will leave to future generations.

Even what appear to be accidental defects are explained by the universal law of optimization of reproduction by natural selection. The naive reader may think that a rotten hole in its trunk is tough luck on a tree, but the informed evolutionary biologist assures us that it is an evolutionarily favored ploy by the tree to attract squirrels who will then spread the tree's seeds far and wide. There is no adversity whose use has not been sweetened by an appeal to natural selection.

The explanation of all of biological phenomena, from the molecular to the social, as special cases of a few over-arching laws is the culmination of a program for the mechanization of living phenomena that began in the seventeenth century with the publication in 1628 of William Harvey's *Exercitatio de motu cordis et sanguinis in animalibus* (On the motion of the heart and blood in animals), in which the circulation of the blood is explained in terms of a mechanical pump with a series of pipes and valves. Descartes' elaboration of a general machine meta-phor for organisms, in Part V of the *Discours*, made exten-sive use of the work of Harvey, to whom he refers with characteristic French hauteur, only as "a physician from England, to whom one must give high praise for having broken the ice in this area." But the machine metaphor creates a general program for biological investigation that is circumscribed by just those properties that organisms have in common with machines, objects that have articu-lated parts whose motions are designed to carry out par-ticular functions. So the program of mechanistic biology has been to describe the bits and pieces of the machine, to show how the pieces fit together and move to make the machine as a whole work, and to discern the tasks for which the machine is designed.

That program has had extraordinary success. We know the structure of living organisms down to the finest details of the internal structure of cells and the folding of molecules, although there remain some important questions open, such as an adequate description of the connections in the brains of large organisms like us. We also understand a great deal about the functions of organs, tissues, cells, and of a remarkably large number of the molecules that make us up. Nor is there any reason to suppose that what is still unknown will not be revealed by the same techniques and with the same concepts that have characterized biology for the last three hundred years. The program of Harvey and Descartes to reveal the details of the *bête machine* has worked. The problem is that the machine metaphor leaves something out, and naive mechanistic biology, which is nothing but physics carried on by other means, has tried to cram it all in at the expense of a true picture of nature.

The problems of biology are not only the problems of an accurate description of the structure and function of the machines, but also the problem of their *history*. Organisms have history at two levels. Each one of us began life as a single fertilized egg cell which underwent processes of growth and transformation to produce a subscriber to *The New York Review of Books*. Those processes will continue and we will be continuously transformed, changing the shape of our bodies and minds, until we end "this strange eventful history." In addition to their individual life stories, organisms have a collective history that started three billion years ago with rudimentary agglomerations of molecules, that has now reached its halfway point with tens of millions of diverse species, and that will end three billion years from now when the Sun consumes the Earth in a fiery expansion. Of course, machines too have histories, but a knowledge either of the history of technology or of the building of individual machines is not an essential part of the understanding of their workings. The designers of modern cars do not have to consult Daimler's original design for an internal combustion engine nor does my garage mechanic need to know how an automobile assembly plant works. In contrast, a complete understanding of organisms cannot be separated from their histories. So the problem of how the brain functions in perception and memory is precisely the problem of how the neural connections come to be formed in the first place under the influence of sights, sounds, caresses, and blows.

The recognition of the historical nature of biological processes is not new. The problem of bringing the individual and collective histories of organisms into one grand mechanical synthesis already represented an important set of questions for eighteenth-century biology and for the Encyclopedists.[3] The biology of the nineteenth century was consumed with the issue, and the two great monuments of biology of the last century were the Darwinian scheme for evolution and the elaboration of experimental embryology by the German school of *Entwicklungsmechanik*.

The fundamental difficulty of fitting these phenomena into the mechanical synthesis arises from an inconvenient property of historical processes, namely their *contingency*. That is, systems in which history is important are systems in which influences outside the structures themselves play an important role in determining their function, so to the extent that those outside forces may vary, the history of the system itself will vary. One does not need to take Tolstoy's extreme anarchic position to agree that the outcome of the battle of Borodino was not determined by the birth either of Napoleon or Kutuzov, nor by the disposition of their troops on September 7, 1812. Any consideration of historical events necessarily demands that we confront the relation between the system that is our object of study and the penumbra of circumstances in which it is embedded, what is inside and what is outside. The relationship between inside and outside is not at issue for the machine, except that what is outside may interfere with its normal functioning. Changes in temperature and violent movements of the base on which it stands are disturbances of the proper motions of a clock, which is why the Admiralty offered a considerable prize for the design of an accurate ship's chronometer. The project to include the life histories of organisms in the machine model then requires that the interaction between the inside and the outside be dealt with and somehow disposed of without compromising the determinist Cartesian program. The embryologists and the evolutionists have taken two quite different approaches to the interaction between the inside and outside, which solve the problem of creating disciplines "of very high generality like that of the physical sciences" but at the expense of seriously distorting our view of living nature and of preventing, in the end, the solution of the very problems that these sciences have set themselves.

The technical word for the process of continual change during the lifetime of an organism is *development*, whose very etymology reveals the theory that underlies its study. Literally, "development" is an unfolding or unrolling, a metaphor that is more transparent in its Spanish equivalent, *desarrollo*, and in the German *Entwicklung*, an unwinding. In this view, the history of an organism is the unfolding and revelation of an already immanent structure, just as when we develop a photograph, we reveal the image that is already latent in the exposed film. The process is entirely internal to the organism, the role of the external world being only a provision of a hospitable condition in which the internal process can run its normal course. At most, some special external condition, say the temperature rising above some minimum, may be necessary to trigger the developmental process, which then unfolds by its own internal logic, as the latent picture becomes manifest when the film is immersed in developing solution.

A characteristic of development theories, whether of the body or the psyche, is that they are *stage* theories. The organism is seen as going through a series of ordered stages, the successful completion of the previous stage being the condition for the initiation of the next. The classical descriptions of animal embryology are in terms of discrete stages, the "two-cell stage," the "four-cell stage," the "blastula [ball of cells] stage," the "neurula [nerve crest] stage." There is then the possibility of arrested development with the system becoming stuck at an intermediate stage, unable to complete its normal life cycle because of an internal fault in the machinery or because the external world has thrown a monkey wrench into the works. Theories of psychic development are classic stage theories. Children must pass successfully through the successive Piagetian stages if they are to understand how to cope with the world of real external phenomena. Freudian theory supposes that abnormality is a consequence of fixation at anal or oral stages on the way to normal genital eroticism. For all these theories, the external world can only trigger or inhibit the normal orderly unfolding of an internally programmed sequence. Thus, developmental biology and psychology finesse the problem of the interplay between inside and outside by denying to the external any creative role.

The claim for the hegemony of internal over external forces in development has been an intellectual commitment since the beginning of developmental biology. Struggles over competing theories of embryogenesis have been carried on entirely within that world view. The most famous was the debate, at the end of the eighteenth and beginning of the nineteenth century, between preformationism and epigenesis.[4] Preformationists, in a view that strikes us as medieval superstition, held that the adult organism was, in fact, already present in minuscule, a homunculus within the fertilized egg (indeed, within the sperm), and that the process of development consisted only in the growth and solidification of the tiny transparent miniature.

Epigeneticists, whose view prevailed in modern biology, claimed that only an ideal plan of the adult existed in the egg, a blueprint that was made manifest by the process of organism building. Except that we now identify that plan with physical entities, the genes made of DNA, nothing much has changed in the theory in the last two hundred years. Yet, between a concrete preformationism that thought there was a little man in every sperm, and an idealist preformationism that sees the complete specification of the adult already present in the fertilized egg, waiting only to be made manifest, there is not much difference except for the mechanical details. In the claim made by one of the world's leading molecular biologists, a co-discoverer of the genetic code, at the centenary observance of Darwin's death, that if he had a large enough computer and the complete DNA sequence of an organism he could compute the organism, we hear the echoes of the eighteenth century. The trouble with the metaphor of "development" is that it gives an impoverished picture of the actual determination of the life history of organisms. Development is not simply the realization of an internal program; it is not an unfolding. The outside matters.

First, even when organisms have a few clearly differentiated "stages," these do not necessarily follow each other in some predetermined order, but the organism, in its lifetime, may pass among the stages repeatedly, depending upon signals from outside. Tropical vines that grow in the deep forest begin life as a germinating seed on the forest floor. In the first stage of growth the vine is *positively* geotropic and *negatively* phototropic. That is, it hugs the ground and grows away from the light toward the dark. This has the effect of bringing the vine to the base of a tree. On encountering a tree trunk, the vine becomes *negatively* geotropic and *positively* phototropic, like most plants, and grows upward along the tree trunk toward the light. In this stage it begins to put out leaves of a characteristic shape. When it gets higher in the tree, where the light intensity is greater, the leaf shape and distance between successive leaves change and flowers appear. Yet higher up, the growing tip of the vine moves out laterally along a branch, again changing its leaf shape, and then changes back to being positively geotropic and negatively phototropic, drops off the branch, and starts to grow straight down toward the forest floor. If it hits another branch, lower down, it starts again an intermediate stage, but if it reaches the ground, it begins its cycle again from the beginning. Depending upon the light intensity and height above the ground, the vine makes different transitions between stages.

Second, the development of most organisms is a consequence of a unique interaction between their internal state and the external milieu. At every moment in the life history of an organism there is contingency of development such that the next step is dependent on the current state of the organism and the environmental signals that are impinging on it. Simply, the organism is a unique result of both its genes and the temporal sequence of environments through which it has passed, and there is no way of knowing in advance, from the DNA sequence, what the organism will look like, except in general terms. In any sequence of environments that we know of, lions give birth to lions and lambs to lambs, but all lions are not alike.

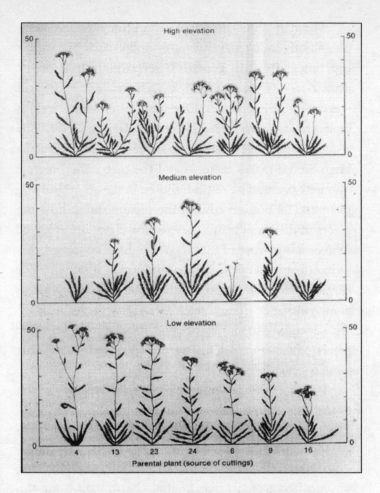

Norms of reaction to elevation for seven different Achillea *plants
(seven different genotypes). A cutting from each plant was grown at low,
medium, and high elevations.* (Carnegie Institution of Washington)

The consequence of this contingency for the variation among individual organisms is illustrated by a classic experiment in plant genetics.[5] Seven individuals of the plant *Achillea* were collected in California and each plant was cut into three pieces. One piece from each plant was replanted at low elevation (30 meters above sea level), one at intermediate elevation (1,400 meters), and one in the High Sierras (3,050 meters), and the pieces then regrew into new plants. The result is shown in the accompanying picture. The bottom row of the picture shows how the pieces of the seven plants grew at low elevation, arranged in decreasing order of their final height. The second row shows the pieces of the same plants grown at intermediate elevation, and the top row is the result of growing pieces at high elevation. The three plants in any vertical column are genetically identical, because they grew from three pieces of the same original plant and therefore carry the same genes.

It is clear that we cannot predict the relative growth of the different plants when the environment is changed. The tallest plant at the low elevation has the *poorest* growth at the intermediate elevation and even fails to flower there. The second largest at the high elevation (plant 9) is intermediate in height at the intermediate elevation, but is the second *smallest* at low elevation. Taken as a whole, there is simply no predictability from one environment to the next. There is no "best" or "largest" genetic type. While we cannot cut people into bits and regrow them in different environments, in every experimental organism where it is possible to duplicate the genetic constitution and test the resultant individuals in different environments, the general result is like that for *Achillea*.

The interaction between genes and environment does not exhaust the sources of variation in development. All "symmetrical" organisms develop asymmetries that fluctuate in direction from individual to individual. The fingerprints of the left and right hands of any individual human being can be distinguished. A fruit fly, no larger than the tip of a lead pencil, having developed while stuck to the inside of a glass culture vessel, has different numbers of sensory bristles on its left and right sides, some flies having more on the left, some more on the right. Moreover, this side-to-side variation is as large as the difference among different flies. But the genes on the left and right sides of a fly are the same, and it seems absurd to think that the temperature, humidity, or concentration of oxygen was different between left and right sides of the tiny developing insect. The variation between sides is a result of random events in the timing of division and movement of the individual cells that produce the bristles, so-called developmental noise. Such noise is a universal feature in cell division and movement and certainly plays a role in the development of our brains. Indeed, one influential theory of central nervous system development puts the random growth and hooking up of nerve cells at the base of the entire process.[6] We simply do not know how much of the difference in cognitive function between different human beings is a consequence of genetic difference, how much is the result of different life experiences, and how much is the result of random developmental noise. I cannot play the viola like Pinchas Zuckerman and I am in serious doubt that I could have done so had I started at the age of five. He and I have different nerve connections, and some of these differences were present at birth, but that is not a demonstration that we are genetically different in this respect.

Despite the evidence of environmental and random variation that is lying about at every hand, developmental biology as a science makes considerable progress holding on to the metaphor of unfolding by restricting the ambit of problems that it addresses to just those that can ignore the external and the indeterminate. Developmental biologists concentrate entirely on how the front end of an animal is differentiated from the rear end and why pigs don't have wings, problems that can indeed be approached from the inside of the organism and which concern some general properties of the machinery. Since the production of "conceptual schemes of high generality" is the mark of success of a science, what biologist will step off the high road to Stockholm to wallow in the slough of individual variation? So the limitations of our conceptual schemes dictate not only the form of our answers to questions but which questions are allowed to be "interesting."

The greatest triumph of nineteenth-century biology was, of course, the elaboration of a mechanistic and materialistic explanation for the history of all of life. The word "evolution" has the same roots as "development" and signifies, literally, an unrolling of an already immanent history. Indeed, some pre-Darwinian theories corresponded to the metaphor, the most influential being Karl Ernst von

Baer's fusion of embryology and evolution in his notion of *recapitulation*. In this scheme, advanced organisms, in their individual development, pass through a series of stages corresponding to the adults of their less evolved ancestral forms. That is, their development recapitulates their evolutionary history. Progressive evolution thus consists in the adding of new stages, but every species will pass through all the old ones on its way from egg to adult. It is indeed true that at an early embryonic stage we have gill slits like fish, connections between the sides of our heart like amphibians, and tails like puppy dogs, all of which disappear as we mature, so we certainly do carry in our individual histories the traces of our evolution.

Darwinian theory made a radical break with this internalist view. Darwin accepted fully the contingency of evolution and constructed a theory in which both internal and external forces play a role, but in an asymmetrical and alienated way. The first step in the theory is the complete causal separation between the internal and the external. In Lamarckism, with its commitment to the inheritance of acquired characteristics and the incorporation of the external into the organism as a consequence of the organism's own strivings, there is no clear separation of what is inside from what is outside. Darwin's radical difference from Lamarckism was his clear demarcation of inside and outside, of organism and environment, and his alienation of the forces within organisms from the forces governing their outside world. According to Darwinism, there are mechanisms entirely internal to organisms that cause them to vary one from another in their heritable characteristics. In modern terms, these are mutations of the genes that control development. These variations are not induced by the environment but are produced at random with respect to the exigencies of the outside world. Quite independently, there is an outside world constructed by autonomous forces outside the influence of the organism itself that set the conditions for the species' survival and reproduction. The inside and outside confront each other only through the selective process of differential survival and reproduction of those organic forms that best match *by chance* the autonomous external world. Those that match survive and reproduce, the rest are cast off. Many are called but few are chosen.

This is the process of *adaptation*, by which the collectivity comes to be characterized by just those forms that by chance fit the preexistent demands of an external nature. Nature poses problems for organisms that they must solve or else perish. Nature, love it or leave it. Again, the metaphor corresponds to the theory. By "adaptation" we mean the altering and tuning of an object to fit some preexistent situation, as when traveling Americans use an adapter to make their razors and hair driers work on European voltages. Evolution by adaptation is the evolution of organisms forced by the demands of an autonomous external world to solve problems that are not of their own making, and their only hope is that the internal force of random mutation will, by chance, provide a solution. The organism thus becomes the passive nexus of internal and external forces. It seems almost not to be an actor in its own history.

Darwin's alienation of the environment from the organism was a necessary step in the mechanization of biology, replacing the mystical interpenetration of interior and exterior that was without any material basis. But what is a necessary step in the construction of knowledge at one moment becomes an impediment at another. While Lamarck was wrong to believe that organisms could incorporate the outer world into their heredity, Darwin was wrong in asserting the autonomy of the external world. The environment of an organism is not an independent, preexistent set of problems to which organisms must find solutions, for organisms not only solve problems, they create them in the first place. Just as there is no organism without an environment, there is no environment without an organism. "Adaptation" is the wrong metaphor and needs to be replaced by a more appropriate metaphor like "construction."

First, while there is, indeed, an external world that exists independent of any living creature, the totality of that world should not be confused with an organism's environment. Organisms by their life activities determine what is relevant to them. They assemble their environments from the juxtaposition of bits and pieces of the outside world. Just outside my window are patches of dry grass, surrounding a large stone. Phoebes gather the grass to make nests in the rafters of my porch, but the stone is not relevant to them and is not part of their environment. The stone, on the other hand, is part of the environment of thrushes, who use it as an anvil to break open snails by rapping them sharply. Not far away is a tree with a large hole in it that is part of the environment of a woodpecker who makes a nest in it, but the hole does not exist in the biological world of the phoebe or thrush. Biologists' descriptions of the "ecological niche" of an organism, say, a bird, have a revealing rhetoric. "The bird," they say, "eats flying insects in the spring, but switches to small seeds in the fall. It makes a nest of grass, twigs, and mud about two to three feet above the ground in the fork of a tree, in which it raises three to four chicks. It flies south when days get shorter than twelve hours."[7]

Every word is a description of the life activities of the bird, not of an autonomous external nature. It is, in fact, impossible to judge what the "problems" set by nature are without describing the organism for which these problems are said to exist. In some abstract sense, flying through the air is a potential problem for all organisms, but this problem does not exist for earthworms who, *as a consequence of the genes they carry*, spend their lives underground. Therefore, just as the information needed to specify an organism is not contained entirely in its genes, but also in its environment, so the environmental problems of the organism are a consequence of its genes. Penguins, birds who spend much of their lives underwater, have altered their wings to make them into flippers. At what stage in the evolution of the flying ancestors of penguins did swimming underwater become a "problem" to be solved? We do not know, but presumably their ancestors had already to have made swimming an important part of their life activities before natural selection could favor turning wings into paddles. Fish gotta swim and birds gotta fly. Nor is the origin of flight without its problems. A flightless animal that sprouted rudimentary wings would get no lift from them at all, as one can easily verify by flapping a pair of ping-pong paddles. Lift force increases very slowly with the increase of surface area for small wings, and below a certain size there is no lift at all. On the other hand, even small thin membranes that can be moved around turn out to be excellent devices for dissipating heat, or collecting it from sunshine, and many butterflies use their wings for that purpose. Our present guess is that wings did not originate to solve the problem of flight at all, but were heat-regulatory devices that, when they became large enough, gave the insect some lift and so made flight a new problem to be solved.

Because organisms create their own environments we cannot characterize the environment except in the presence of the organism that it surrounds. Using appropriate optical devices it is possible to see that there is a layer of warm moist air surrounding each one of us which moves continually up the surface of our bodies and off the tops of our heads. This layer, present in all organisms that live in air, is a result of the production of heat and water by our metabolism. As a consequence we carry around with us our own atmosphere. If the wind should blow and strip away that boundary layer we would be exposed to the world outside and only then would we know how cold it really is out there. That is the meaning of the wind-chill factor.

The attempt to define an environment in the absence of organisms can cost the taxpayers and embarrass scientists. The Mars Lander, among its other projects, carried a device to detect life on Mars. One suggestion for this device was to have a long sticky tongue that would unroll, pick up dust, and then retract to bring the dust under a microscope whose images could be sent back to Earth. This is the morphological test for life. If it has a suggestive shape or it wiggles, it is alive. This design was rejected in favor of a physiological test.

The Mars Lander carried a vacuum cleaner filled with radioactive soup. A hose would suck up dust into the soup and detectors would sense the production of radioactive carbon dioxide as the Martian organisms metabolized the soup. And indeed, carbon dioxide was produced to the ecstatic, although brief, wonder of the observers on Earth. Unfortunately, the production of carbon dioxide stopped rather suddenly in a way that is not characteristic of aging bacterial cultures, and after much discussion it was decided that there was, after all, no life on Mars, and that the observations were the result of chemical reactions on the surface of the dust. The problem of the experiment was that it defined life on Mars by creating an environment for that life based on terrestrial environments. But earthly environments are the product of earthly organisms. Until we know what Martian organisms (if any) are like, we will not know how to construct environments to trap and detect them.

Second, every organism, not just the human species, is in the constant process of changing its environment, both creating and destroying its own means of subsistence. It is part of the ideology of the environmental movement that alone among species, human beings are in the process of destroying the world that they inhabit, and that undisturbed nature is in unchanging harmony and balance. There is nothing here but Rousseauian romance. Every species consumes its own resources of space and nutriment, and in the process produces waste products that are toxic to itself and its offspring. Every act of consumption is an act of production and every act of production an act of consumption. Every animal, when it breathes in precious oxygen, exhales poisonous carbon dioxide, poisonous to itself, but not to plants, who thrive on it. As Mort Sahl once observed, no matter how cruel and unfeeling we may be, every time we breathe we make a flower happy. Every organism deprives its fellows of space and, when it feeds and digests, excretes toxic waste products into its own neighborhood.

In some cases as a matter of their normal function, organisms make it impossible for their own offspring to succeed them. When the stony farms of New England were abandoned in the westward rush after 1840, the untilled fields were at first occupied by herby weeds and then were taken over by pure stands of white pine. In the early 1900s it was thought that the pines would be a steady source of income from wood and pulp, but they failed to replace themselves and gave way to hardwoods, immediately when cut and slowly when left alone. The problem is that pine seedlings are intolerant of shade and cannot grow up in a forest, even a forest of pines. The adult pines create a condition that is inimical to their own offspring, so that they can survive as a species only if some of the seeds can, like the eighteenth- and nineteenth-century children of European farmers, colonize newly opened areas where they are not oppressed by their parents. But all organisms also produce the conditions necessary for their existence. Birds make nests, bees hives, and moles burrows. When plants put down roots they change the texture of the soil and excrete chemicals that encourage the growth of symbiotic fungi which help the plant's nutrition. Fungus-gardening ants gather and chew up leaves, which they seed with the spores of mushrooms that they eat. At every moment every species is in the process of creating and re-creating, both beneficially and detrimentally, its own conditions of existence, its own environment.

It may be objected that some important elements of the outer world are thrust on organisms by the very laws of nature. After all, gravitation would be a fact of nature even if Newton had never existed. But the relevance to an organism of external forces, even of gravitation, is coded in its genes. We are oppressed by gravity, acquiring flat feet and bad backs by virtue of our large size and upright posture, both consequences of the genes we have inherited. Bacteria, living in a liquid medium, do not experience gravity, but they are subject to another "universal" physical force, Brownian motion. Because they are so small, bacteria are buffeted about by the random thermal motions of molecules in the liquid medium, a force which, fortunately, does not send us reeling from one side of the room to the other. All natural forces operate effectively in particular ranges of size and distance so that organisms, as they grow and evolve, may move from the domain of one set of forces to another. All the organisms that now exist have evolved and must survive in an atmosphere that is 18 percent oxygen, an extremely reactive and chemically powerful element. But the earliest organisms did not have to cope with free oxygen, which was absent from the aboriginal atmosphere, an atmosphere with high concentrations of carbon dioxide. It is organisms themselves that have produced the oxygen, through photosynthesis, and have depleted the carbon dioxide to the fraction of a percent that it now represents by trapping it in vast deposits of limestone, coal, and petroleum. The proper view of evolution is then of co-evolution of organisms and their environments, each change in an organism being both the cause and the effect of changes in the environment. The inside and the outside do indeed interpenetrate and the organism is both the product and location of that interaction.

The constructionist view of organism and environment is of some consequence to human action. A rational environmental movement cannot be built on the demand to save the environment, which, in any case, does not exist. Clearly, one does not want to live in a world that smells and looks worse than at present, in which life is even more solitary, poor, nasty, brutish, and short than it now is. But that wish cannot be realized by the impossible demand that human beings stop changing the world. Remaking the world is the universal property of living organisms and is inextricably bound up in their nature. Rather, we must decide what kind of a world we want to live in and then try to manage the processes of change as best we can to approximate it.

Notes

[1] The original study in the late 1940s was by C. C. North and P. K. Hatt. "Jobs and Occupations: A Popular Evaluation," in L. Wilson and W. L. Kolb, *Sociological Analysis* (Harcourt Brace, 1949). Later studies gave essentially identical results.

[2] Free Press, 1952, p. 14.

[3] In Diderot's *Le Rêve de d'Alembert*, both the physician Bordeu's waking expositions and d'Alembert's sleeping meanderings center on these issues. "Who knows what races of animals preceded us? Who knows what races of animals will succeed ours?" d'Alembert wonders in his sleep. "How could we know that the man, leaning on his stick, whose eyes are blind, who drags himself along with such effort, yet more different on the inside than the outside, is the same man who yesterday walked so lightly, shifting with ease the heaviest loads...?" Bordeu asks. These biological questions were such a central concern of philosophy that Diderot has Mlle de l'Espinasse remark sarcastically that "there is no difference between a physician who is awake and a philosopher who is asleep."

[4] For a perceptive and informative view of this debate see Shirley Roe, *Matter, Life and Generation: Eighteenth-Century Embryology and the Haller-Wolff Debate*. (Cambridge University Press, 1981).

[5] J. Clausen, D. D. Keck, and W. W. Hiesey, "Environmental Responses of Climatic Races of *Achillea*," Carnegie Institution of Washington Publication 581, 1958.

[6] The selective theory of the formation of the central nervous system is explicated in G. Edelman, *Neural Darwinism: The Theory of Neuronal Group Selection* (Basic Books, 1989).

[7] The sophisticated bird-watcher will not recognize any real bird in this composite life history.

Oliver Sacks, M.D.

Scotoma: Forgetting and Neglect in Science

I

We may look at the history of ideas backward or forward—we can trace the earlier stages, the intimations, and the anticipations of what we think now; or we can concentrate on the evolution, the effects and influences of what we once thought. Either way, we may imagine that history will be revealed as a continuum, an advance, an opening like the tree of life. What one often finds, however, is very far from a majestic unfolding, and very far from being a continuum in any sense. This is a conclusion that I will try to illustrate by some stories (which might be multiplied a hundredfold) of how odd, complex, contradictory, and irrational the processes of scientific discovery can be. And yet, beyond the twists and anachronisms in the history of science, beyond the vicissitudes and fortuities, perhaps there is an overall pattern to be discerned.

I began to realize how elusive scientific history can be when I became involved with my first love, chemistry. I vividly remember, as a boy, reading a history of chemistry by F. P. Armitage, a former master at my school, and learning that oxygen had been all but discovered in the

1670s by John Mayow, along with a theory of combustion and respiration. But Mayow's work was then forgotten and concealed from view by a century of obscurantism (and the preposterous phlogiston theory), and oxygen was only rediscovered a hundred years later, by Lavoisier. Mayow died at thirty-four: "Had he lived but a little longer," Armitage adds, "it can scarcely be doubted that he would have forestalled the revolutionary work of Lavoisier, and stifled the theory of phlogiston at its birth." Was this a romantic exaltation of John Mayow, or a romantic misreading of the structure of the scientific enterprise, or could the history of chemistry have been wholly different, as Armitage suggests?[1]

I thought of this history in the mid-Sixties, when I was a young neurologist just starting work in a headache clinic. My job was to make a diagnosis—migraine, tension headache, whatever—and prescribe treatment. But I could never confine myself to this, nor could many of the patients I saw. They would often tell me, or I would observe, other phenomena: sometimes distressing, sometimes intriguing, but not strictly part of the medical picture—not needed, at least, to make a diagnosis.

Often in a classical migraine there is an aura, so-called, where the patient may see scintillating zigzags slowly traversing the field of vision. These are well described and understood. But sometimes, more rarely, patients would tell me of more complex geometrical patterns that appeared in place of, or in addition to, the zigzags: lattices, whorls, funnels, and webs, all shifting, gyrating, and modulating constantly. When I searched the current literature, I could find no mention of these. Puzzled, I decided to go back and look at nineteenth-century accounts, which tend to be much fuller, much more vivid, much richer in description, than modern ones.

143

My first discovery was in the rare book section of our college library (everything written before 1900 counted as "rare")—an extraordinary book on migraine written by a Victorian physician, Edward Liveing, in the 1860s. It had a wonderful, lengthy title: *On Megrim, Sick-Headache, and Some Allied Disorders: A Contribution to the Pathology of Nerve Storms*, and it was a grand, meandering sort of book, clearly written in an age far more leisurely, less rigidly constrained, than ours. It touched briefly on the complex geometrical patterns I had been told of, and it referred me to a paper written a few years before, "On Sensorial Vision," by John Frederick Herschel, son of Frederick Herschel (both father and son, as well as being eminent astronomers, had "visual" migraines and wrote about them). I felt I had struck paydirt at last. The younger Herschel gave meticulous, elaborate descriptions of exactly the phenomena my patients had described; he had experienced them himself, and he ventured some deep speculations about their possible nature and origin. He thought they might represent "a sort of kaleidoscopic power" in the sensorium, a primitive, pre-personal generating power in the mind, the earliest stages, even precursors, of perception.

I could find no adequate description of these "Geometrical Spectra," as Herschel called them, in the entire hundred-year period between his observations and my own, and yet it was clear to me that at least one person in twenty affected with migraine experienced them on occasion. How had these phenomena—startling, highly characteristic, unmistakable hallucinatory patterns—evaded notice for so long? In the first place, someone must make an observation and report. In the same year that Herschel reported his spectres, G. B. A. Duchenne, in France, described a case of muscular dystrophy. But here the stories diverge. As soon as Duchenne's observations were published, physicians started "seeing" the dystrophy everywhere, and within a few years, scores of further cases were reported and described. The disorder had always existed, ubiquitous and unmistakable. Why did we need Duchenne to open our eyes? His observations entered the mainstream of clinical perception at once, as a syndrome, a disorder of great importance.

Herschel's paper, by contrast, sank without a trace. He was not a physician making medical observations but an independent observer of great curiosity. He considered himself an astronomer even in regard to his own hallucinations, and indeed called himself "an astronomer of the inward." Herschel suspected that his observations had scientific importance, that such phenomena could lead to deep insights about the brain, but whether they had medical importance too was not in his mind. Since migraine was usually defined as a "medical" condition, Herschel's observations had no professional status; they were seen as irrelevant, and after a brief mention in Liveing's book were forgotten, ignored by the profession. If they were to point to new scientific ideas about the mind and brain, there was no way of making the connection in the 1850s; the necessary concepts only emerged 120 years later.

These necessary concepts emerged in conjunction with the recent development of chaos theory, which shows that while it is impossible to predict in detail the individual disposition of each element in a system, when there are a large number of elements in interaction (as, for example, with the million-odd nerve cells in the primary visual cortex), patterns can be discerned at a higher level by using recently developed methods of mathematical and computer analyses. There are "universal behaviors" which emerge in such interactions, behaviors which represent the ways such dynamic, nonlinear systems organize themselves. They tend to take the form of complex, reiterative patterns in space and time—indeed the very sort of networks, whorls, spirals, and webs that one sees in the geometrical hallucinations of migraine.

Such chaotic behaviors have now been recognized in a vast range of natural systems, from the eccentric motions of Pluto to the striking patterns that appear in the course of certain chemical reactions, to the multiplication of slime fungi and the vagaries of the weather. With this, a hitherto insignificant or unregarded phenomenon like the geometrical patterns of migraine aura suddenly assumes a new importance. It shows us, in the form of a hallucinatory display, not only an elemental activity of the cerebral cortex, but an entire self-organizing system, a universal behavior, at work.[2]

II

With migraine, I had to go back to an earlier, forgotten medical literature—a literature that most of my colleagues saw as superseded or obsolete. I also found myself in a similar position with Tourette's syndrome, the "*maladie des tics*" described in the 1880s by Georges Gilles de la Tourette. My interest in it had been kindled in 1969 when I was able to "awaken" a number of *encephalitis lethargica* patients with L-DOPA and saw how many of them rapidly swung from motionless, trancelike states through a tantalizingly brief "normality" and then to the opposite extreme—violently hyperkinetic tic-ridden states very similar to the half-mythical "Tourette's syndrome." I say "half-mythical" because no one in the 1960s spoke much about Tourette's; it was considered extremely rare and possibly factitious. I had only vaguely heard of it. Things were soon to change: in the 1970s, Tourette's was rediscovered, and found to be a thousand times commoner than suspected; there was a surge of interest in the syndrome and research into its course.

But this surge of interest, this rediscovery, followed a silence and neglect of sixty years or more, during which the syndrome was rarely discussed or even diagnosed. Indeed, when I started to think about it, in 1969, as my own patients were becoming palpably Tourettic, I had difficulty finding any current references whatever, and once again had to go back to the literature of the previous century: to Gilles de la Tourette's original papers in 1885 and 1886 and to the dozen or so reports that followed them. It was an era of superb, mostly French, descriptions of the varieties of tic behavior, which culminated (and terminated) in the book *Tics* published in 1902 by Henry Meige and E. Feindel. Yet between 1903 and 1970, the syndrome itself seemed almost to have disappeared.

Why? One must wonder whether this neglect was not caused by the growing pressures at the beginning of the new century to try to explain scientific phenomena, following a time when it was enough to *describe* them. And Tourette's was peculiarly difficult to explain. In its most complex forms it could express itself not only as convulsive movements and noises but as tics, compulsions, obsessions, and tendencies to make jokes and puns, to play with boundaries and engage in social provocations and elaborate fantasies. Though there were attempts to explain the syndrome in psychoanalytical terms, these, while casting light on some of the phenomena, were impotent to explain others; there were clearly organic components as well. In 1960, the finding that a drug, haloperidol, which counters the effects of dopamine, could extinguish many of the phenomena of Tourette's generated a much more tractable hypothesis—that Tourette's was, essentially, a chemical disease, caused by an excess of (or an excessive sensitivity to) the neurotransmitter dopamine in the brain.

With this comfortable, reductive explanation to hand, the syndrome suddenly sprang into prominence again, and indeed seemed to multiply its incidence a thousand-fold. There is now a very intensive investigation of Tourette's syndrome, but it is an investigation almost confined to its molecular and genetic aspects. And while these may explain some of the overall excitability of Tourette's, they may do little to illuminate the particular forms of the Tourettic disposition to engage in comedy, fantasy, mimicry, mockery, dream, exhibition, provocation, and play. Thus, while we have moved from an era of pure description to one of active investigation and explanation, Tourette's itself has been fragmented in the process, and is no longer seen as a whole.[3]

This sort of fragmentation is perhaps typical of a certain stage in science—the stage that follows pure description. But the fragments must somehow, sometime, be gathered together, and presented once more as a coherent whole. This will require an understanding of determinants at *every* level, from the neurophysiological to the psychological to the sociological—and of their continuous and intricate interaction.[4]

III

I had spent fifteen years as a physician making neurological observations, but in 1974 I had a neurological experience of my own—experienced, so to speak, the "inside" of a neuropsychological syndrome. I severely injured the nerves and muscles of my left leg while climbing in a remote part of Norway. I needed surgery to connect the muscle tendons and time to allow the healing of nerves. During the two-week period in which the leg was denervated and immobilized in a cast, it was not only bereft of movement and sensation, it ceased to feel like a part of me. It seemed to have become a lifeless, almost inorganic object, not real, not mine, inconceivably alien and strange. But when I tried to communicate the experience to my surgeon, he said, "Sacks, you're unique. I've never heard of anything like this from a patient before."

I found this absurd. How could I be "unique"? There must be other cases, I thought, even if my surgeon had not heard of them. As soon as I was mobile enough, I started to talk to my fellow patients, and many of them, I found, had similar experience of "alien" limbs. Some had found the experience so uncanny and fearful that they had tried to put it out of their minds; others had worried about it secretly, but not tried to communicate it.

When I left the hospital, I went to the library determined to seek out the literature on the subject. For three years I found nothing. Then I came across an account by Silas Weir Mitchell, the great nineteenth-century American neurologist, fully and carefully describing phantom limbs ("sensory ghosts," as he called them). Weir Mitchell also wrote of "negative phantoms," experiences of the subjective annihilation and alienation of limbs following severe injury and surgery. He encountered vast numbers of such cases during the Civil War, and was so struck by them that he had at once published a special circular on the matter ("Reflex Paralysis"), which was distributed by the Surgeon General's office in 1864. His observations aroused brief interest, then disappeared. More than fifty years had to pass before the syndrome was rediscovered. This again occurred during wartime, when thousands of new cases of neurological trauma were seen at the front. In 1917, the eminent neurologist J. Babinski published (with J. Froment) a monograph entitled *Syndrome Physiopathique*, in which, apparently ignorant of Weir Mitchell's report, he described the syndrome I had. Once again the observations sank without a trace. (When, in 1975, I finally came upon the book in our library, I found I was the first person to have borrowed it since 1918.) During World War II, the syndrome was fully and richly described for the third time by two Soviet neurologists, A. N. Leont'ev and A. V. Zaporozhets, again in ignorance of their predecessors. Though their book, *Rehabilitation of the Hand*, was translated into English (in 1960), their observations completely failed to enter the consciousness of either neurologists or rehabilitation specialists.

As I pieced together this extraordinary, even bizarre story, I felt more sympathy with my surgeon and his saying he had never heard of anything like my symptoms before. And yet the syndrome is not that uncommon: it occurs whenever there is a significant dissolution of body image. But why is it so difficult to put this on record, and to give the syndrome its due place in our neurological knowledge and consciousness?

The term "scotoma" (darkness, shadow)—as used by neurologists—denotes a disconnection or hiatus in perception, essentially a gap in consciousness produced by a neurological lesion. Such lesions may be at any level, from the peripheral nerves, as in my own case, to the sensory cortex of the brain. It is therefore extremely difficult for a patient with such a scotoma to be able to communicate what is happening. He himself, so to speak, scotomizes the experience. It is equally difficult for his physician and his listeners to take in what he is saying, for they, in turn, tend to scotomize what they are hearing. Such a scotoma is literally unimaginable unless one is actually experiencing it (this is why I suggest, only half jocularly, that people read *A Leg to Stand On* while under spinal anesthesia, so that they will know in their own persons what I am talking about).[5]

If, somehow, by an almost superhuman effort, these barriers to communication are transcended—as they were by Weir Mitchell, Babinski, and Leont'ev and Zaporozhets—no one seems to read or remember what they have written. There is a historical or cultural scotoma, a "memory hole," as Orwell would say.

IV

Let us return from this uncanny realm to a more positive (but still strangely neglected or scotomized) phenomenon, in particular that of total colorblindness following a cerebral injury or lesion—a so-called acquired cerebral achromatopsia. (This is, of course, a completely different condition from common colorblindness, which is caused by a deficiency of one or more color receptors in the retina.) I select this example because I have explored it in some detail, but I learned of it quite by accident, when a patient with the condition wrote to me, asking if I had ever encountered it before. My friend and colleague Dr. Robert Wasserman and I spent a great deal of time with this extraordinary patient, and our original report appeared in *The New York Review of Books* in 1987.[6]

But when we looked into the history of this condition, we soon encountered a remarkable gap or anachronism. Acquired cerebral achromatopsia—and even more dramatically, hemiachromatopsia, the loss of color perception in only one half of the visual field, coming on suddenly as a consequence of a stroke—had been described in exemplary fashion by a Swiss neurologist, Louis Verrey, in 1888. When his patient subsequently died, Verrey was able to delineate the exact area of the visual cortex which had been damaged by her stroke. Here, he said, "the center for chromatic sense will be found." Within a few years of Verrey's report, there were other careful reports of similar problems with color perception and the lesions that caused them, and achromatopsia and its neural basis seemed firmly established. But then, strangely, there were no more reports—not a single full case report for seventy-five years, between the last nineteenth-century report in 1899 and the "rediscovery" of achromatopsia in 1974.

This story has been discussed, with great scholarship and acumen, by two colleagues of mine: Semir Zeki of the University of London and Antonio Damasio of the University of Iowa. Zeki, remarking that resistance to Louis Verrey's findings started the instant they were published, sees their virtual denial and dismissal as springing from a deep and perhaps unconscious philosophical attitude, the then prevailing belief in the seamlessness of vision. The notion that we are given the visual world as a datum, an image, complete with color, form, movement, and depth, is a natural and intuitive one, which had seemingly been given scientific and philosophical legitimation by Newtonian optics and Lockean sensationalism. The invention of the camera lucida, and later of photography, seemed to exemplify such a mechanical model of perception. Why should the brain behave any differently? Color, it was obvious, was an integral part of the visual image, and not to be dissociated from it. The ideas of an isolated loss of heretofore normal color perception and of a center for chromatic sensation in the brain were thought self-evident nonsense. Verrey had to be wrong; such absurd notions had to be dismissed out of hand. So they were, and achromatopsia "disappeared."

Darwin often remarked that no man could be a good observer unless he was an active theorizer as well, and Darwin himself, his son Francis wrote, seemed "charged with a theorizing power" that would animate and illuminate all his observations, even the most trivial ones. But, Francis added, this power was always balanced by skepticism and caution, and above all by experiments that as often as not demolished the new theory. Theory, though, can be a great enemy of honest observation and thought as well, especially when it forgets that it *is* theory or model and hardens into unstated, perhaps unconscious, dogma or assumption. Mistaken assumptions killed Verrey's observation, killed the entire subject for three quarters of a century.[7]

The notion of perception as "given" in some seamless, overall way was finally shaken to its foundations by the findings of David Hubel and Torsten Wiesel in the late Fifties and early Sixties that there were cells and columns of cells in the visual cortex which acted as "feature detectors," specifically sensitive to horizontals, verticals, edges, alignments, and other features of the visual field. The idea began to develop that vision had components, that visual representations were in no sense "given," like optical images or photographs, but were *constructed* by an enormously complex and intricate correlation of different processes. Perception was now seen as composite, as modular; the interaction of a huge number of components. The seamlessness of perception was not "given," but had to be *achieved* in the brain.

It thus became clear in the 1960s that vision was an analytic process, depending on the differing sensitivities of a large number of cerebral (and retinal) systems, each "tuned" to respond to different components of perception. It was in this atmosphere of hospitality to subsystems and their integration that Zeki discovered specific cells sensitive to wavelength and color in the visual cortex of the monkey; and he found them in much the same area as Verrey had suggested as a color center eighty-five years before. Zeki's discovery seemed to release clinical neurologists from their almost century-long inhibition. Within a few years, scores of new cases of achromatopsia were found, and it was at last legitimized as a valid neurological condition.[8]

V

The subject of color has been endlessly fascinating to artists, scientists, and philosophers alike. Spinoza's first treatise, written when he was nineteen, was on the rainbow. Newton wrote about "the celebrated phaenomenon of Colours," and, on the basis of his famous prism experiments, concluded that white light was composite, and that the color of its constituent rays was determined by their differing "refrangibilities."

155

The idea of a purely physical determination of color, however, was anathema to Goethe when he started his own explorations of it.[9] Reality, for him, was not to be found in the simplifications and idealizations of physics but in the complex phenomenal reality of experience. Intensely aware of the subjective reality of colored shadows and colored afterimages, of the effects of contiguity and illumination on the appearance of colors, he felt that these, rather than a Newtonian prism in a dark room or a spectrum on a wall, must be the basis of a proper color theory. Goethe was fascinated above all by the subjectivity of color, and its unexpected appearances and modifications and disappearances, which seemed to resist physical explanation. It was evident to him that colors were constructed by the mind in an exceedingly complex manner, not at all comparable to a simple physical reproduction.

From the moment in 1790 when he seized a prism and cried, "Newton is wrong!" Goethe set himself to disproving the Newtonian hypothesis (as he understood or misunderstood it) and to constructing his own color theory. In the last forty years of his life, color became an obsession for Goethe, and he regarded his color studies, his *Farbenlehre*, as fully as important as his entire poetic opus. It would be for this, he hoped, that he might be remembered, when *Faust* had long since passed into oblivion.

Classical color theory was undisturbed in spite of all the reservations and fulminations of Goethe. His color theory was seen by his scientific contemporaries as unscientific and mystical, and this was not helped by the contemptuous way in which he referred to Newton and others. His style and language were alien to those of contemporary scientific researchers. And there was a feeling, steadily increasing after 1800, that poets and scientists were set apart, had their own place, and that Goethe was trespassing in a realm not his own.

A century and a half passed—and then, in 1957, something singular happened. Edwin Land (already famous for his invention of the Polaroid instant camera, but also an experimenter and theorist of great audacity, indeed genius) staged a demonstration that stunned everybody who saw it and was wholly inexplicable according to classical color theory.

Newton had shown that if one mixed colored lights (for example, orange and yellow), one obtained something intermediate (an orange-yellow). Almost three hundred years later, Land repeated this, but he used colored lights to project black-and-white transparencies of a still life taken through filters of these same colors. If only the yellow beam was used, one saw a monochromatic, yellow still life; if only the orange beam was used, a similar, orange monochrome. When both beams were turned on, the audience expected something intermediate, but what they saw instead was a sudden bursting into full color, with reds, blues, greens, purples, every color in the original still life. Impossible! An illusion!

It is indeed an illusion, but such an illusion as Goethe considered colored shadows to be—the sort of illusion which prompted him to say, "Optical illusion is optical truth!" He was intensely aware that there was not any simple equivalence of wavelength and color (as Newton thought), and felt that color was not a simple sensation but an "inference" or "act of judgment." Feeling this intuitively, but wholly ignorant of what physiological mechanisms could allow such an inference, Goethe made a great error: he bypassed physiology, made a mystical leap to "the mind," and proposed an entire mental or subjective theory (or pseudo-theory) of color.

What Land was able to do was to "save the phenomena," to re-explore the very real phenomena which so fascinated Goethe, while giving them an objective explanation such as Goethe could not (not least because the needed advances in physiology and psychophysics were only made after his death). Thus Land showed that colors are not perceived in isolation, a scene is not an aggregate of colored points; there is, instead, a complete surveying of the scene as a whole, and a minute comparison and correlation of the sensations from each part. This involves, first, an extraction of "color-separation" images, records of the lightness and darkness of each part of the scene as transmitted by the three different color receptors in the retina, and then a comparison of these three "lightness records" in the brain.

It is the brain's act of comparing these three records that forms the basis of our inference or judgment of color. Thus it is only now, with the dazzling demonstrations of Land and the physiological work of Zeki (who has shown the area in the brain where such "inferences" are made) that Goethe's insights have been confirmed. None of this contemporary work, however, makes any reference to Goethe, and the *Farbenlehre* remains as unread as it was in 1832.

VI

Can we draw any lessons from the examples I have been discussing? I believe we can. One might first invoke the concept of prematurity here—and see the nineteenth-century observations of Goethe, Herschel, Weir Mitchell, Tourette, and Verrey as having come before their times, so that they could not be integrated into contemporary conceptions. Gunther Stent, writing about "prematurity" in scientific discovery, says, "A discovery is premature if its implications cannot be connected by a series of simple logical steps to canonical, or generally accepted, knowledge." He discusses this in relation to the classical case of Mendel, and the lesser known but fascinating case of Oswald Avery (who discovered DNA in 1944—a discovery totally overlooked at the time, because of a lack of knowledge that would have enabled scientists to appreciate its importance).[10] Prematurity, I think, though relatively rare in science, may be much commoner in medicine, partly because medicine does not have to do elaborate experiments (as Mendel or Avery did) but, in the first place, can simply describe.[11]

But "scotoma" involves more than prematurity, it involves the *deletion* of what was originally perceived, a loss of knowledge, a loss of insight, a forgetting of insights that once seemed clearly established, a regression to less perceptive explanations. All these not only beset neurology but are surprisingly common in all fields of science. They raise the deepest questions about why such lapses occur. What makes an observation or a new idea acceptable, discussable, memorable? What may prevent it from being so, despite its clear importance and value?

Freud would answer this question by emphasizing resistance: the new idea is deeply threatening or repugnant, and hence is denied full access to the mind. This doubtless is often true, but it reduces everything to psychodynamics and motivation; and even in psychiatry, this is not enough.

It is not enough to apprehend something, to "get" something, in a flash. The mind must be able to accommodate it, to retain it. This process of accommodation, of being able to create a mental space, a category with potential connections—and the readiness to do this— seems to me crucial in determining whether an idea or discovery will take hold and bear fruit, or whether it will be forgotten, fade, and die without issue. The first difficulty, the first barrier, lies in one's own mind, in allowing oneself to encounter new ideas and then to bring them into full and stable consciousness, and to give them conceptual form, holding them in mind even if they do not fit, or contradict, one's existing concepts, beliefs, or categories. Darwin remarks on the importance of "negative instances" or "exceptions," and how crucial it is to make immediate note of them, for otherwise they are "sure to be forgotten."

That it is crucially important to take note of exceptions, and not forget them or dismiss them as trivial, was brought out in "On Unnoticed Sensations and Errors of Judgement," the first paper written by Wolfgang Köhler, before his pioneer work in Gestalt psychology. Köhler spoke here of premature simplifications and systemizations in science, psychology in particular, and how they could blind one, ossify science, and prevent its vital growth.

"Each science," he wrote, "has a sort of attic into which things are almost automatically pushed that cannot be used at the moment, that do not quite fit.... We are constantly putting aside, unused, a wealth of valuable material [which leads to] the blocking of scientific progress." (1913)

Thus, at the time Köhler wrote this, visual illusions were seen as "errors of judgment"—trivial, of no relevance to the workings of the mind-brain. But Köhler would soon show that the opposite was the case, that such illusions constituted the clearest evidence that perception does not just passively "process" sensory stimuli but actively creates large configurations or "gestalts" which organize the entire perceptual field. These insights now lie at the heart of our present understanding of the brain as dynamic and constructive. But it was first necessary to seize on an "anomaly," a phenomenon contrary to the accepted frame of reference, and by according it attention, to enlarge it, to revolutionize it entirely.

But if anomalies promise a transition to a larger mental space, they may do so through a very painful, even terrifying, process of undermining one's existing beliefs and theories—painful because our mental lives are sustained, consciously or unconsciously, by theories, sometimes invested with the force of ideology or delusion.

The history of science and medicine, in a sort of Darwinian way, has taken much of its shape from intellectual and personal competitions that force people to confront both anomalies and deeply held ideologies; and competition, in the form of open debate and trial, is therefore essential to its progress. When these debates are open and straightforward, a rapid resolution and advance can sometimes be achieved.[12]

This is "clean" science, but there is a good deal of "dirty" science too. A most unfortunate instance of it occurred when the astrophysicist S. Chandrasekar, as a teenager of genius, developed a startling theory of stellar degeneration. While stars below a certain critical mass might become "white dwarfs," those of greater mass, he argued, would instead show "relativistic degeneracy," and implode into another state of being altogether. The great theoretical astronomer A. S. Eddington, who had been his patron and protector, attacked him and his theory with extraordinary virulence, and prevented him from developing and publishing it. (Eddington's attack, as became apparent later, was based on a mistaken, indeed delusional, cosmological theory, a house of cards which, he felt, would collapse if Chandrasekar's theory was true.) The result was that theoretical astronomy was held back for thirty years, and a theory of black holes (so clearly implied by Chandrasekar's thinking in the early 1930s) was fully worked out only in the late 1960s.[13]

Attack by authority, in this way, can have the direst consequences, not only preventing the publication or dissemination of crucial observations and thoughts, but sometimes resulting in the destruction of the victim. Chandrasekar was a man of much resilience and psychic toughness. When attacked by Eddington he fought for a while, and then, when this proved hopeless, turned his attention elsewhere, and did fundamental work in other fields. For Georg Cantor, the great mathematician who developed the ideas of transfinite cardinals and infinite sets, the outcome was not so auspicious: persecuted by the renowned Felix Klein, whom he called "the great field-marshal of German mathematics," he became floridly psychotic (though he continued to develop his theory in lucid periods in between). And Ludwig Boltzmann, the supreme theoretical physicist of the closing decades of the nineteenth century, was driven to suicide by misunderstanding and attack; had he lived only a very little longer, he would have seen the worldwide recognition of his methods and ideas of statistical mechanics.

Two of the greatest authorities on color perception during the nineteenth century—Hermann von Helmholtz and Ewald Hering—proposed diametrically different theories: Helmholtz argued that color was composed by the adding together of the responses from the three color receptors on the retina, whereas Hering thought that there was a more complicated mechanism, which involved competition and inhibition between the activities of the three types of cone, and a relating of this process to the retina's capacity to respond to brightness. Disagreement spread from this particular issue to virtually every open question in visual psychophysics and physiology. An intense and bitter personal rancor between the two men extended to their students and followers for fifty years after their deaths. Steven Turner, in a recent account of this, ascribes the conflict to "perceptual incommensurability" between the two; but C.R. Cavonius, in his review of Turner's book, puts it down to "sheer bloody-mindedness."[14]

In extreme cases scientific debate can threaten to destroy the belief systems of one of the antagonists, and with this, perhaps, the belief system of the entire culture. The clearest example of such destruction followed the publication of *Origin of Species* in 1859, with the furious debates between "Science" and "Religion" (embodied in the conflict between Thomas Huxley and Bishop Wilberforce), and the violent but pathetic rearguard actions of Agassiz, who felt his lifework, his sense of a creator, massacred by Darwin's theory.[15]

In this extreme case, the anxiety of obliteration was such that Agassiz actually went to the Galápagos himself and tried to duplicate Darwin's experience and findings, in order to repudiate the theory which he felt had annihilated him.

It has often surprised me similarly that chaos theory was not discovered or invented by Newton or Galileo— they must have been perfectly familiar, for example, with the phenomena of turbulence and eddies which are constantly seen in daily life (and so consummately portrayed by Leonardo). Perhaps they avoided thinking of their implication, foreseeing these as potential infractions of a rational, lawful, orderly Nature.

This indeed is much what Poincaré felt when, more than two centuries later, he became the first to investigate the mathematical consequences of chaos: "These things are so bizarre that I cannot bear to contemplate them." Now we find the patterns of chaos beautiful—a new dimension of nature's beauty—but this was certainly not how it seemed to Poincaré.

The most famous example of such a repugnance in our own century is, of course, Einstein's violent distaste for the seemingly irrational nature of quantum mechanics. Even though he himself had been one of the very first to demonstrate quantum processes (and indeed received the Nobel Prize for this), he refused to consider quantum mechanics as anything more than a superficial representation of natural processes, which would give way, with deeper insight, to a more harmonious and orderly one.

Darwin was at pains to say that he had no forerunners, that the idea of evolution was not in the air; Newton, despite his famous comment about "standing on the shoulders of giants," also denied such forerunners. This "anxiety of influence" (which Harold Bloom has discussed powerfully in regard to the history of poetry) is a potent force in the history of science as well. One may have to believe that others are wrong; one may have to, as Bloom insists, misunderstand others, and (perhaps unconsciously) react against others, in order to successfully develop and unfold one's own ideas. ("Every talent," Nietzsche writes, "must unfold itself in fighting.") In the case of Hering and Helmholtz, it seems to me neither could bear the idea of being significantly influenced by the other; they had to split apart, to deny each other, to insist (with a sort of desperation) on their own originality. Each behaved as if the world did not have room for them both; each felt threatened by the very existence of the other, as if the rightness of one, on any issue, annihilated the other.[16] Now, more than a century later, we can see that both men were partly right (in their theories, not in the ways they fought each other over them), that their research and ideas were not antagonistic but complementary, and that an adequate theory of color vision must draw equally upon them both.[17]

VII

We have spoken of discoveries or ideas so "premature" as to be almost without connection or context, hence unintelligible, or ignored, at the time; and of others passionately, at times ferociously, contested, in the necessary but often brutal *agon* of science. But perhaps it is equally important to look at discoveries and ideas not only by taking account of their acceptance or dismissal by contemporaries but by seeing them with respect to the history of ideas, as some of the greatest scientists have done.

Einstein entitled his own idiosyncratic book *The Evolution of Physics*,[18] and the story he tells is not just one of emergence, but one of radical discontinuity too. Thus Part I is entitled "Rise of the Mechanical View," and Part II "Decline of the Mechanical View." The mechanical world view, as he sees it, had to collapse and leave a rather frightening intellectual vacuum before a radically new concept could be born. The conception of a "field" of forces—which was a prerequisite of the theory of relativity—in no way emerges or evolves *from* a mechanical one. Thus it is less evolution than revolution that Einstein speaks of—a revolution which he himself, of course, took to heretofore unimaginable heights.

But, most importantly, Einstein is at pains to say that the new theory does not destroy the old, does not invalidate it, or supersede it, but, rather, "...allows us to regain our old concepts from a higher level." Einstein expands this notion in a famous simile:

> To use a comparison, we could say that creating a new theory is not like destroying an old barn and erecting a skyscraper in its place. It is rather like climbing a mountain, gaining new and wider views, discovering unexpected connections between our starting point and its rich environment. But the point from which we started out still exists and can be seen, although it appears smaller and forms a tiny part of our broad view gained by the mastery of the obstacles on our adventurous way up.

Helmholtz, in his partly autobiographical memoir, *Thought in Medicine*, also uses the image of a mountain climb (he himself was an ardent alpinist), but describes the climb as anything but linear. One cannot see in advance, he says, a way up the mountain; it can only be climbed by trial and error. The intellectual mountaineer makes false starts, gets stuck, gets into blind alleys and cul-de-sacs, finds himself in untenable positions, has to backtrack, has to descend and to start again. Thus, slowly and painfully, with innumerable errors and corrections, he makes his zigzag way up the mountain. It is only when he reaches the summit or the height he desires that he will see that there was, in fact, a direct route, a "royal road," to it. In his publications, Helmholtz says, he takes his readers along this royal road, but this bears no resemblance to the crooked and tortuous processes by which he constructed a path for himself.

167

In such accounts we find a common theme—that there is some vision, intuitive and inchoate, of what must be done, and that it is this, once glimpsed, which drives the intellect forward. Thus Einstein at fifteen had fantasies about riding a light beam and ten years later developed the theory of special relativity, going from a boy's dream to the grandest of theories. Was the achievement of the theory of special relativity, and then of general relativity, "inevitable," part of an ongoing historical process? Or the result of a singularity, the advent of a unique genius? Would relativity have been conceived in Einstein's absence? (And how quickly would relativity have been accepted had it not been for the solar eclipse of 1917, which, by a rare chance, allowed the theory to be confirmed by accurate observation of the effect of the sun's gravity on light? One senses the fortuitous here no less than inevitability—and, not trivially, a requisite level of technology, one which could measure Mercury's orbit accurately.) Neither "historical process" nor "genius" is an adequate explanation—each glosses over the complexity, the chancy nature, of reality. What emerges from a close study of a life such as Einstein's is the immense role that fortuity played in his life and the fact that this or that technical achievement was available to be used—the Michelson-Morley experiment, for example. If Riemann and other mathematicians had not developed non-Euclidean geometries, Einstein would not have had the intellectual techniques available to move from a vague vision to a fully developed theory, which requires the concepts of non-Euclidean geometries. He was, of course, intensely alert, primed to perceive and seize whatever he could use. "Chance favors the prepared mind" (Claude Bernard), but it was a particularly happy coincidence that non-Euclidean geometries had been developed at this time. They had been worked out as pure abstract constructions, with no notion that they might be appropriate to any physical model of the world.

A huge number of isolated, autonomous individual factors must be present before the seemingly magical act of a creative advance, and the absence (or insufficient development) of any one may suffice to prevent it. The huge role of contingency, of sheer luck (good or bad), it seems to me, is never emphasized enough. And this is much more so in medicine even than in science, for medicine often depends crucially on rare and unusual, perhaps unique, cases being encountered by the "right" person, at the right time.

Cases of prodigious memory are naturally rare, and the Russian Shereshevsky was among the most remarkable of these. But he would merely be remembered now (if at all) as "another case of prodigious memory" had it not been for the chance of meeting A. R. Luria, himself a prodigy of clinical observation and insight. It required the genius of a Luria, and a thirty-year-long exploration of Shereshevsky's mental processes, to produce the unique insights of Luria's great book, *The Mind of a Mnemonist.* Hysteria, by contrast, is not uncommon, and has been well described since the eighteenth century. But it was not plumbed psychodynamically until a brilliant, articulate hysteric encountered the original genius of the young Freud and his friend Breuer. Would psychoanalysis, one wonders, ever have got going without Anna O.—and the specially receptive, prepared minds of Freud and Breuer, without the intersection of a unique subject and explorer? (I am sure that it would have, but later, and in a different way.)

Thus there is both fortuity and inevitability in science. If Watson and Crick had not cracked the double helix of DNA in 1953, Linus Pauling would almost certainly have done so. The structure of DNA, one might say, was ready to be discovered—though who did it, and how, and exactly when, remained unpredictable. In the seventeenth century, similarly, the time was ripe for the invention of calculus, and this was devised by both Newton and Leibniz, almost simultaneously, though in entirely different ways.

Could the history of science—like life—be rerun quite differently? Does the evolution of ideas resemble the evolution of life? Assuredly we see sudden explosions of activity, when enormous advances are made in a very short time—this was so, for molecular biology, in the 1950s and 1960s; for quantum physics, in the 1920s; and a similar burst of fundamental work seems to be occurring in neuroscience now, in the "Decade of the Brain." Sudden bursts of discovery change the face of science, and these are often followed by long periods of consolidation, and, in a sense, stasis. One cannot but be reminded of the picture of "punctuated equilibrium" given us by Niles Eldridge and Stephen Jay Gould, and wonder if there is at least an analogy here to a natural evolutionary process.

And yet, even if this holds as a general pattern, the specifics, one feels, could be very different, for ideas seem to arise, flourish, go in all directions, or abort and become extinct, in completely unpredictable ways. Gould is fond of saying that if the tape of life on earth could be replayed, it would be wholly different the second time around. Suppose that John Mayow had indeed discovered oxygen in the 1670s; that Babbage's Difference Engine—a computer—had been built in the last century; might the course of science have been quite different?[19] This is the stuff of fantasy, of course, but fantasy which brings home the sense that science is not an ineluctable process but, in its details, contingent in the extreme.

VIII

And yet, are there, beneath the contingencies, underlying patterns, stages through which science and medicine must go? Descriptions of diseases—pathographies—already refined in antiquity, became wonderfully informative in the middle and latter part of the nineteenth century, when thousands of distinct and distinctive neurological disorders were described in a detail that has never been surpassed. There was a huge openness to experience at this time, a love for phenomena, a genius for describing them, with a sort of cartographic passion for classifying and mapping them—but often with little or no thought about their nature or "meaning."[20]

There was, of course, a good deal of neurological theorizing in the late nineteenth century, too—indeed the greatest of neurological theorists, Hughlings Jackson, made his most important contributions at this time. But, on the whole, description flourished despite this theorizing, and in a sort of detachment from it, as if minute phenomenological description needed nothing beyond itself—an attitude epitomized in Goethe's maxim: "Everything factual is, in a sense, theory.... There is no sense in looking for something behind phenomena: they *are* theory."

Indeed in this era of naturalistic description and phenomenological passion for detail, a concrete habit of mind seemed highly appropriate, and an abstract or ratiocinating one suspect—an attitude beautifully brought out by William James in his famous essay on Agassiz:

The only man he really loved and had use for was the man who could bring him facts. To see facts, not to argue or reason, was what life meant for him; and I think he often positively loathed the ratiocinating type of mind.... The extreme rigor of his devotion to this concrete method of learning was the natural consequence of his own peculiar type of intellect, in which the capacity for abstraction and causal reasoning and tracing chains of consequences from hypotheses was so much less developed than the genius for acquaintance with vast volumes of detail, and for seizing upon analogies and relations of the more proximate and concrete kind.

(Pascal, at the very start of the *Pensées*, speaks of two fundamentally different types of mind: the intuitive and the mathematical—Agassiz's mind, it is clear, was almost overwhelmingly of the first type.)

The great glory of descriptive neurology, then, came in the latter half of the nineteenth century, and it is to these descriptions that we must turn and return, even now, if we are to be reminded of the fullness of the phenomena, and a sense of the whole (just as I myself found when I discovered Liveing's *Megrim*).

But then, with the turn of the century, something seemed to go out of medicine: the great descriptions, and great describers, which had been its glory, seemed to vanish. And with the ending of this tradition, there came over medicine a certain sense of loss, an amnesia. A. R. Luria, in his autobiography, speaks poignantly of this.

A parallel development is described by William James, and is epitomized for him in the rise and fall of Agassiz. He describes how the young Agassiz, coming to Harvard in the mid-1840s, "studied the geology and fauna of a continent, trained a generation of zoologists, founded one of the chief museums of the world, gave a new impulse to scientific education in America"—and all this through his passionate love of phenomena and facts, of fossils and living forms, his passionate and lyrical concreteness of mind, his scientific and religious sense of a divine system, a whole. But then, James intimates, there came a transformation: zoology itself was losing its old form, changing from a natural history, intent on wholes—species and forms and their taxonomic relationships—to studies in physiology, histology, chemistry, pharmacology, a new science of the micro, of mechanisms and parts abstracted from a sense of the organism, and its organization as a whole. This was a transformation to which Agassiz's mind, the intuitive mind, could not well adapt, and which, therefore, pushed him away from the center of scientific thinking and made him, in his later years, an eccentric and tragic figure. [21]

Nothing was more exciting, more potent than this new science—and yet it was clear that something was being lost, too. As James remarks, writing in 1896:

> In the fifty years that have sped since [Agassiz] arrived here, our knowledge of Nature has penetrated into joints and recesses which his vision never pierced. The causal elements and not the totals are what we are now most passionately concerned to understand; and naked and poverty-stricken enough do the stripped-out elements and forces occasionally appear to us.

This mixed sense of gain and loss has been felt by neurologists since the founding of their discipline in the 1860s— the discovery in 1862 by Broca that a specific language defect, aphasia, could be ascribed to damage of a specific part of the brain. The part Broca pointed to was promptly conceived as a "language area" of the brain, and its discovery led the way to the delineation of scores, and ultimately hundreds, of other areas, each (it was believed) a center for a form of behavior or perception, a particular neural or psychic function. By the turn of the century, it was felt, this allocation of functions was approaching completeness, the brain, the mind, now being seen as composite in nature, a vast mosaic of different functions all lodged in separate centers of their own; and with this a sense of the whole was gone.

And yet there was also, within the ranks of neurologists, a division between those, a majority, who favored such a "parcellation" of brain activities and others—preeminently Hughlings Jackson, Pierre Marie, Henry Head, and Kurt Goldstein—who felt that there must be higher-order functions (such as memory, attention, emotion, thought, consciousness, identity) which required large-scale or global processes in the brain, and could not be adequately understood in terms of small, discrete areas of the brain. (Neurologists themselves, in a sort of jocular slang, referred to localizers as "splitters," and to globalists or holists as "lumpers.")

Such a history, of course, is almost a caricature—no figure of note has ever been a pure splitter or lumper, and virtually all have struggled, in different ways, to reconcile the two approaches. James himself, more conscious than anybody else of both the peculiar strengths and weaknesses of the two approaches, envisaged a third stage which would transcend both and reconcile them:

> The truth of things is after all their living fulness, and
> some day, from a more commanding point of view
> than was possible to anyone in Agassiz's generation, our
> descendants, enriched with the spoils of all our analytic
> investigations, will get round again to that higher and
> simpler way of looking at Nature.

Such a "higher and simpler way" was yearned for by the
would-be holists of earlier generations, but it was beyond
their grasp, beyond the reach of any concepts available in
their time. But, until recently, scientists inclined to holism
had little more to go on than a feeling of a tantalizing
hiatus between the sense of nature (including human nature) given to us by direct observation and intuition and
the fragmented, decentered sense that science provided.[22]

IX

During the past forty-five years, across the entire range of
sciences, from cosmology to geology to embryology to
neurology, new conceptions are emerging that could not
have been arrived at without drawing on the spoils of our
analytic investigations, the century or more of "micro"
science which preceded them, but which also do not de-
rive directly from these—any more than the physicist's
concept of "field" is derivable from a mechanical world
view.

These new concepts, typically, must be synthetic in
nature, expressing general principles of large-scale, global
organization, which bring unity to the seemingly

fragmented observations of micro-science. The search for and the emergence of such general principles are especially active now in astronomy and cosmology—as with the Big Bang theory and superstring theory—and in nuclear physics, where the present dizzying array of "elementary particles" is awaiting the confirmation of their place in the so-called standard model. Beyond these are hopes for theories that will unite all the fundamental forces of nature.

In biology, and especially neurobiology and neuropsychology, the matter is even more complex, because there are many levels of organismic function, from the molecular to the ontological, to be considered in relation to one another, and brought, ideally, into a single synthetic theory. For this reason there may need to be partial general principles—general principles applying to particular levels—before anything beyond these can be envisaged. No such concepts became available until the 1930s, when P. K. Anokhin and other Soviet theorists developed the notion of a "functional system." Thus where classical neurology had thought in terms of functions and centers, one now has to think in terms of complex dynamic systems adapted to particular biological ends, in which all the intermediate links could be varied within wide limits. At a higher level, a crucial concept developed by Elkhonon Goldberg, a pupil of Luria, is that of cortical and cognitive gradients.

In "lower" animals, and "lower" parts of the brain, there is a "hard wiring" of neurological function—everything (or almost everything) from respiratory function to instinctual responses is genetically determined, and assigned to fixed nuclei or modules in the brain. This may also be the case in the primary sensory areas (for example, the areas where color and motion are "constructed"). But at higher levels, in the association areas, Goldberg argues, where learning occurs, an entirely new principle of organization comes into being. These areas, by contrast, are uncommitted at birth, and their development depends on the particularities of life experience: they *assume* a function in the course of life. This higher cortex is conceived as forming a continuum, a sheet or network or field, with different densities, or "gradients," of representations across them; and its functions, in Goldberg's words, are

> continuous, interactive, and emergent, as opposed to mosaic, modular, and prededicated.

When Goldberg's hypothesis of cortical gradients was first broached, almost twenty years ago, the idea (as Hume said of his own *Treatise*) "fell deadborn from the press." There was no reaction in the neurological literature; the concept did not become part of the neurological consciousness. The idea was too novel, too strange, too "premature," to be taken in, at a time when everybody thought in terms of modules, of a cortical mosaic consisting of small areas, each identified with particular functions. Now, twenty years later, the atmosphere in neurology has radically changed—not least through forms of synthetic neural modeling which show what interactive neural networks can achieve—and has become very

sympathetic to the idea of self-organization, of cerebral
and mental organization *emerging* under the influence of
experience. Goldberg's models, which were dismissed at
their inception, now demand intensive attention.

An even more ambitious and audacious general
theory, in that it attempts to embrace every level of
cerebral function and evolution, is Gerald Edelman's
theory of neuronal group selection, or "neural Darwinism." For here every level, from the micro-patterns of
synaptic weights and neuronal connectivities to the
macro-patterns of an actual lived life, are, in broad principle, brought together and made one.[23] It may be that this
far-reaching theory too is also premature, in that while
some of it fits beautifully with existing clinical and neuro-
scientific knowledge, some of it goes beyond existing
facts, and is not readily testable at present—but the same
might be said of any revolutionary theory.

A new sort of theorizing, or world view, has appeared in the last forty years, and, so to speak, across the
board, from the notions of self-organization in simple
physical and biological systems to notions of complex
adaptive systems in nature and society, from concepts of
emergent properties in neural networks to concepts of
neuronal group selection in the brain. A new "macro"
view has been emerging not only of the integral character
of nature but of nature as innovatory, emergent, creative.

Nietzsche spoke of "the world as a work of art that
gives birth to itself." The very word "nature" connotes
birth, creativity—but this sense disappeared with the rise
of the Newtonian and thermodynamic world views,

which (in the words of Paul Davies, the cosmologist) "present the universe either as a sterile machine, or as a state of degeneration and decay." Now through the power of new syntheses and global theories, this impoverished and, so to speak, lifeless picture of nature is being transcended, so that we are beginning to be able to reapproach the natural world, reappropriate it once more, but this time *through* science, as an emerging, self-creating whole: that "higher and simpler" way of looking at it of which William James dreamed.

Notes

1 Armitage's book was written in 1905, to stimulate the enthusiasm of Edwardian schoolboys, and it seems to me now, with different eyes, that it had a somewhat romantic and jingoistic ring, an insistence that it was the English, not the French, who discovered oxygen. William Brock, in his recent *History of Chemistry*, writes: "Early historians of chemistry liked to find a close resemblance between Mayow's explanation and the later oxygen theory of calcination." But such resemblances, Brock stresses, "are superficial, for Mayow's theory was a mechanical, not a chemical, theory of combustion. [Moreover]...it marked a return to a dualistic world of principles and occult powers." But the same could be said, to some extent, of Boyle—all the greatest innovators of the seventeenth century, not excluding Newton, still had one foot in the medieval world of alchemy, the hermetic and the occult—indeed, Newton's intense interest in alchemy and the esoteric continued to the end of his life. (This was first brought out, startlingly, by J. M. Keynes in his tercentenary essay "Newton, the Man," and the overlap of "modern" and "occult" in the climate of seventeenth-century science is now well accepted.)

2 I described the phenomena of migraine aura in the original (1970) edition of my book *Migraine*, but could only say then that they were "inexplicable" by then existing concepts. I have discussed them in the new light of chaos theory, in collaboration with my colleague Dr. Ralph Siegel, in an additional chapter in the revised (1993) edition.

3 I have written about Tourette's syndrome, and the history of our ideas about it, in "Tics," *The New York Review*, January 29, 1987, pp. 37–41.

4 A somewhat similar sequence has occurred in "medical" psychiatry. If one looks at the charts of patients institutionalized in asylums and state hospitals in the 1920s and 1930s one finds extremely detailed clinical and phenomenological observations, often embedded in narratives of an almost novelistic richness and density (as in the "classical" descriptions of Kraepelin and others at the turn of the century). With the institution of rigid diagnostic criteria and

manuals (the "diagnostic statistical manuals" called DSM-III and DSM-IV) this richness and detail and phenomenological openness have disappeared, and one finds instead meager notes that give no real picture of the patient or his world, but reduce him and his disease to a list of "major" and "minor" diagnostic criteria. Present-day psychiatric charts in hospitals are almost completely devoid of the depth and density of information one finds in the older charts, and will be of little use in helping us to bring about the synthesis of neuroscience with psychiatric knowledge which we so need. The "old" case histories and charts, however, will be invaluable.

⁵ I have discussed this more fully in the afterword to the revised edition of *A Leg to Stand On*.

⁶ "The Case of the Colorblind Painter" appeared in *The New York Review*, November 19, 1987, pp. 25–34. An expanded and revised version of the article also appears in my book *An Anthropologist on Mars* (Knopf, 1995).

⁷ And yet there seem to have been other factors at work as well. Damasio describes how when the renowned neurologist Gordon Holmes published his findings on two hundred cases of war injuries to the visual cortex in 1919, he stated summarily that none of these showed isolated deficiencies in color perception, and that his research gave no support to any notion of a color center in the brain. Holmes was a man of formidable authority and power in the neurological world, and his empirically based antagonism to the notion of an isolated cerebral color defect or color center, reiterated with increasing force for over thirty years, was a major factor, Damasio feels, in actually preventing the clinical recognition of the syndrome.

⁸ That conceptual bias was responsible for the dismissal and "disappearance" of achromatopsia is confirmed by the completely opposite history of central motion blindness (akinetopsia), which was described, in a single case, by Zihl et al. in 1983. This patient could see people, or cars, at rest, but they disappeared as soon as they moved, and then reappeared, motionless, in different places. Zihl's case, Zeki notes, was "immediately accepted by the neurological ...and neurobiological world, without a murmur of dissent...in contrast to the more turbulent history of achromatopsia." This

dramatic difference relates to timing, to the profound change in intellectual climate which had come about in the years immediately before. In the early 1970s it had been shown that there was a specialized area of motion-sensitive cells in the prestriate cortex of monkeys, and the *idea* of functional specialization was fully accepted within a decade. Thus by 1983, in Zeki's words, "...all conceptual difficulties had been removed." There was no longer any conceptual reason for rejecting Zihl's findings—indeed, quite the contrary: they were embraced with delight, as a superb piece of clinical evidence in consonance with the new climate.

9 Though Goethe showed himself an admirable empirical scientist (for example, in his discovery of the intermaxillary bone), he was also given to mystical pseudo-theories, such as his almost Platonic theory of the archetypal plant (*Ur-Pflanze*). This was, perhaps, part of the romantic reaction against a mechanical world view, Newton's in particular, which seized Europe at this time, and was particularly strong in Germany (in the form of *Naturphilosophie*), and in England, where Blake fulminated about "simple vision and Newton's sleep."

George Henry Lewes, in his deeply sympathetic but not uncritical biography of Goethe, devotes the longest chapter in this to "The Poet as a Man of Science," and ascribes some of Goethe's weaknesses, as well as his enormous strengths, to the same disposition of mind: "The native direction of his mind," Lewes writes, "is visible in his optical studies as decisively as in his poetry; that direction was towards the *concrete* phenomenon, not towards abstractions.... He was continually laughing at the Newtonians about their prisms and Spectra, as if Newtonians were pedants who preferred their dusky rooms to the free breath of heaven. He always spoke of observations made in his garden, or with a simple prism in the sunlight, as if the natural and simple Method was much more certain than the artificial Method of Science. In this he betrayed his misapprehension of Method.... Hence his failure; hence also his success..." (Lewes, *The Life of Goethe* [Frederick Ungar, 1965], pp. 352–353). The intriguing relation of Goethe's concepts to current theories of color is finely discussed by Arthur Zajonc in *Catching the Light: The Entwined History of Light and Mind* (Bantam, 1993).

10 Gunther Stent's article, "Prematurity and Uniqueness in Scientific Discovery," appeared in the *Scientific American* in December 1972.

When I visited W. H. Auden in Oxford, in February 1973, he was greatly excited by Stent's article, and we spent much time discussing it. Auden wrote a lengthy reply to Stent, contrasting the intellectual histories of art and science; this was published in the March 1973 *Scientific American*.

[11] Had Stent been a geneticist, rather than a molecular biologist, he might have recalled the story of the pioneer geneticist Barbara McClintock, who in the 1940s developed a theory—of so-called jumping genes—which was almost unintelligible to her contemporaries. It was not until thirty years later, when the atmosphere in biology had changed, and had become more hospitable to such notions, that McClintock's insights were belatedly recognized as a fundamental contribution to genetics. Fortunately she lived long enough to have her contribution recognized: the work she had done in the 1940s finally received the Nobel Prize in the 1980s.

Had Stent been a geologist, he might have given another famous (or infamous) example of prematurity—Alfred Wegener's theory of continental drift, proposed in 1915, forgotten or derided for many years, but then rediscovered forty years later, with the rise in plate tectonics theory.

Had Stent been a mathematician he might have cited, as an astonishing example of "prematurity," Archimedes' invention of the calculus two thousand years before Newton's and Leibniz's.

And had he been an astronomer, he might have spoken not merely of a forgetting, but of a most momentous regression, in the history of astronomy. A heliocentric picture of the solar system was clearly established by Aristarchus, in the third century B.C., and well understood and accepted by the Greeks (it was further amplified by Archimedes, Hipparchus, and Eratosthenes). Yet in the second century A.D. it was turned on its head by Ptolemy, and replaced by a geocentric picture of almost Babylonian complexity. The Ptolemaic darkness, the scotoma, lasted 1,400 years, until a heliocentric picture was re-established by Copernicus.

[12] If one aspect of science lies in the realm of competition and rivalry, another, much less healthy, one springs from epistemological misunderstanding and schism, often of a very fundamental sort. One such schism is described eloquently by the biologist Edward O. Wilson in his recent autobiography, *Naturalist*, with regard to his own early work in entomology and taxonomy. He found

his work (and his subject) dismissed by James Watson as not real science, and as no more than "stamp-collecting." Such a dismissive attitude was almost universal among molecular biologists in the 1960s. Ecology, similarly, was scarcely allowed status as a "real" science until quite recently, and is still seen as much "softer" than molecular biology, for example. There is perhaps something of a parallel between this and the often dismissive attitude of "hard" science to clinical medicine, and especially case histories. Freud himself commented on this:

> It still strikes me as strange that the case histories I write read like short stories and that, as one might say, they lack the serious stamp of science. I must console myself with the reflection that the nature of the subject is evidently responsible for this, rather than any preference of my own...

And yet (for all the current talk of hermeneutics) it is evident that Freud's case histories, and all deep case histories, *are* serious science, and they embody and point to a science of the individual which is every bit as "hard" as physics or molecular biology.

13 Kameshwar Wali, in his biography of Chandrasekar, writes, "It took nearly three decades before the full significance of [Chandrasekar's] discovery was recognized, and the Chandrasekar limit entered the standard lexicon of physics and astrophysics. Five decades would pass before he was awarded the Nobel Prize."

Chandrasekar himself later said, "It is quite an astonishing fact that someone like Eddington could have such an incredible authority...and incredible...that there were not people who had boldness and understanding enough to come out and say Eddington was wrong.... I personally believe that the whole development of astronomy, of theoretical astronomy, particularly with regard to the evolution of stars and the understanding of the observations relating to white dwarfs, were all delayed by at least two generations because of Eddington's authority." (Cited in Wali, *Chandra: A Biography of S. Chandrasekar*, University of Chicago Press, 1991.)

14 R. Steven Turner, *In the Eye's Mind: Vision and the Helmholtz-Hering Controversy* (Princeton University Press, 1994), reviewed by C. R. Cavonius, "Not Seeing Eye to Eye," in *Nature*, Vol. 370 (July 28, 1994), pp. 259–260.

15 Gosse, at once deeply devout and a great naturalist, was so torn in half by the debate that he was driven to publish an extraordinary book, *Omphalos*, in which he maintained that fossils do not correspond to any creatures that ever lived, but were merely put in the rocks by the Creator to rebuke our curiosity—an argument which had the unusual distinction of infuriating zoologists and theologians in equal measure.

Darwin himself was often appalled by the very mechanism of nature whose workings he saw so clearly. He expresses this in a letter which he wrote to his friend Hooker in 1856: "What a book a Devil's Chaplain might write on the clumsy, wasteful, blundering low and horribly cruel works of nature!"

16 This attitude of murderous envy toward a rival or imagined rival, from which even the most gifted and fortunate may not be exempt, has been discussed by Leonard Shengold in a very recent article, "Envy and Malignant Envy," *The Psychoanalytic Quarterly*, Vol. 58, No. 4, pp. 615–640 (1994).

17 Competition, while it is the very dynamo of progress, may also be highly destructive in the realm of technology (not least because so many technologies have commercial incentives and investments even stronger than their intellectual ones). Thus Edison showed great ruthlessness and dishonesty in dealing with anyone he felt was a competitor, and in a way which often caused technological delay. His insistence on using direct current for electrical transmission, and his violent and unscrupulous attacks on his fellow inventor Nikola Tesla (who had suggested the far more sensible use of alternating current), retarded the expansion of power-transmission technology for two decades. His treatment of Le Prince, the inventor of the motion picture camera (insofar as any single person could be said to have invented so complex a technology), may have delayed the development of motion pictures; certainly it helped Edison secure an enormous share of the profits from these. This is described by Christopher Rawlence in *The Missing Reel* (London: Collins, 1990).

It was similar in the sphere of automobile technology. In the first decade of the twentieth century, internal combustion engines were matched, if not exceeded, by steam and electric ones. Steam cars, unlike locomotives, could develop a head of steam in thirty seconds, and the Stanley Steamer reached 100 mph in 1909. New

York, it might be noted, had electric cabs by 1898. The development of steam cars and electric cars was fatally opposed by gas interests, to the enduring detriment of both travel and our cities.

18 This book was written in collaboration with Einstein's friend and colleague Leopold Infield, but its thoughts and tone are pure Einstein.

19 Some of these as-if fantasies have been powerfully explored by science fiction writers. Thus William Gibson and Bruce Sterling in their 1991 novel, *The Difference Engine*, imagine science and the world set on a different course with the actual construction of Charles Babbage's Difference Engine (and the start of a computer era) in the 1850s. (Intriguingly, Babbage's Analytical and Difference Engines have now been made, exactly as he specified them, and are on display in the Science Museum in London. They work, and could in fact have been made a century and a half ago, although the cost would have been prodigious.)

20 It is said, even now, that if a young neurologist feels he has discovered a new syndrome, he will do well to check W. R. Gowers's 1888 *Manual*, to make sure it was not discovered a century before.

21 A somewhat similar development, it seems to me, occurred in the lifetime of Humphry Davy—the analogous changes of chemistry, one might argue, occurring in the 1820s, not the 1870s. Davy, like Agassiz, was a genius of concreteness and analogical thinking. He lacked the power of abstract generalization which was so strong in his contemporary John Dalton (it is to Dalton that we owe the foundations of atomic theory) and the massive systematic powers of his contemporary Berzelius. Davy hence descended, from his idolized position as "the Newton of chemistry" in 1810, to being almost marginal fifteen years later. The rise of organic chemistry, with Wöhler's synthesis of urea in 1828—a new realm in which Davy had no interest or understanding—immediately started to displace the "old" inorganic chemistry, and added to Davy's sense of being outmoded in his last years.

Jean Amery, in his powerful book *On Aging*, speaks of how tormenting the sense of irrelevance, or obsolescence, may be; in particular the sense of being *intellectually* outmoded, through the rise of new methods, theories, or systems. Such outmoding, in science, can occur almost instantly when there is a major shift of thought, and there is no doubt that Davy, like Agassiz, knew the torment of it to the full.

22 The sense of a Whole, unapprehensible by analytic science, is a potent generator of dualism. Such a split fosters the idea of a "soul" or "spirit" irreducible to physiology. Thus James's close friend F.W.H. Myers, in the opening paragraph of his book *Human Personality*, was driven to the following vision of the mind:

> I regard each man as at once profoundly unitary and al-
> most infinitely composite, as inheriting from earthly an-
> cestors a multiplex and "colonial" organism—polyzoic
> and perhaps polypsychic in an extreme degree; but also as
> ruling and unifying that organism by a soul or spirit ab-
> solutely beyond our present analysis.

Similarly, the great embryologist Hans Driesch, in the 1890s, awed by what seemed to be the unanalyzably complex organization of the developing embryo (the notion of chemical "organizers" had not yet emerged), resorted to the Aristotelian notion of entelechy, which he conceived, in quasi-mystical terms, as an immanent principle or soul unfolding at the cellular level.

Notions of preformed entities (such as souls or entelechies) presuppose another world, a world of ideal eternal forms—above our own world of becoming, of vicissitude and contingency. Though such suppositions would seem to be anti-scientific in the extreme, they may play a crucial part in the genesis of scientific ideas—thus the mystical notion of *harmonia mundi*, a harmony of the spheres, led Kepler, finally, to his laws of planetary motion.

23 Edelman's most recent exposition of his theory is to be found in *Bright Air, Brilliant Fire* (Basic Books, 1993). I have twice written in detail about Edelman's thought, its relation to the history of our ideas of brain and mind, and its implications for neuroscience, clinical neurology, and developmental psychology, today. *The New York Review*, November 22, 1990, p. 44, and *The New York Review*, April 8, 1993, p. 42.

Notes on Contributors

Stephen Jay Gould teaches geology, biology, and the
history of science at Harvard University. His many books
include *Ever Since Darwin: Reflections in Natural History;
The Panda's Thumb,* which won the American Book
Award; *The Mismeasure of Man; The Flamingo's Smile;
Time's Arrow, Time's Cycle;* and, most recently, *Eight Little
Piggies: Reflections in Natural History.*

Daniel J. Kevles heads the Program in Science, Ethics, and
Public Policy at the California Institute of Technology.
He has written widely on the development of science
and its relationship to society past and present. His
works include *The Physicists: The History of a Scientific
Community in Modern America* and *In the Name of Eugenics:
Genetics and the Uses of Human Heredity,* and he is
co-editor of *The Code of Codes: Scientific and Social Issues
in the Human Genome Project.*

R. C. Lewontin is a leading geneticist and the author of
Biology as Ideology: The Doctrine of DNA and *The Genetic
Basis of Evolutionary Change,* and co-author of *The
Dialectical Biologist* (with Richard Levins) and *Not in Our
Genes* (with Steven Rose and Leon Kamin). He is
Professor of Population Sciences and the Alexander
Agassiz Professor of Zoology and Professor of Biology at
Harvard University.

Jonathan Miller is a doctor of medicine and a director of theater, television, and opera. He is the author of *The Body in Question* (on the history of medicine) and *States of Mind* (on madness), both based on his BBC television series of the same names; *The Human Body*; *The Facts of Life*; and *Subsequent Performances*. He has staged productions at opera houses around the world, including the English National Opera, the Metropolitan Opera in New York, and Glimmerglass. His theater work includes many Shakespeare productions, mainly for London's Royal National Theatre. Dr. Miller has been a Research Fellow in Neuro-psychology at the University of Sussex.

Oliver Sacks is Professor of Neurology at the Albert Einstein College of Medicine. His books include *Migraine, Awakenings, A Leg to Stand On, The Man Who Mistook His Wife for a Hat, Seeing Voices*, and, most recently, *An Anthropologist on Mars: Seven Paradoxical Tales*. The film *Awakenings* and Peter Brook's play *The Man Who* were both based on his work.

Sources of Illustrations

Every effort has been made to secure permission for the illustrations published herein. We gratefully acknowledge the following individuals and institutions for their assistance.

Page 4: Mesmer, after a woodcut by Figuier, courtesy of UPI/ Bettmann, New York.

Pages 46 and 47: (Figure 1) Used with permission of the Toshiba Corporation.

Page 48: (Figure 2) From *The New Yorker*, July 30, 1990. Used with permission of The New Yorker Magazine, Inc.

Page 52: (Figure 3) From *Earth Before the Deluge* by Louis Figuier, 1863, fig. 42: "Vue idéale de la terre pendant la période dévonienne." Courtesy of the Science and Technology Research Section, Science, Industry and Business Library, The New York Public Library, Astor, Lenox and Tilden Foundations.

Page 54: (Figure 4) From *Parade of Life Through the Ages*, *National Geographic* magazine, February 1942. Used with permission of Charles Knight/National Geographic Society Image Collection.

Page 56: (Figure 5) From *Earth Before the Deluge* by Louis Figuier, 1863, fig. 196: "L'Iguanodon et le Megalosaure." Courtesy of the Science and Technology Research Section, Science, Industry and Business Library, The New York Public Library, Astor, Lenox and Tilden Foundations.

Page 57: (Figure 6) From *Parade of Life Through the Ages*, *National Geographic* magazine, February 1942. Used with permission of Charles Knight/National Geographic Society Image Collection.

Page 58: (Figure 7) From *Earth Before the Deluge*, "Appearance of Man," revised version, sixth edition, by Louis Figuier, 1867.

For further information about Granta Books
and a full list of titles, please write to us at

Granta Books

2/3 HANOVER YARD

NOEL ROAD

LONDON

N1 8BE

enclosing a stamped, addressed envelope

———————————

You can visit our website at

http://www.granta.com